ALSO BY JAIME S. CARVALHO

Life: its physics and dynamics

FROM STARS TO SOCIETIES

A PROCESS IN TIME

FROM STARS TO SOCIETIES

A PROCESS IN TIME

Jaime S. Carvalho

American Literary Press
Five Star Special Edition
Baltimore, Maryland

From Stars to Societies: A Process in Time

Library of Congress
Cataloging in Publication Data
ISBN-13: 978-1-934696-32-3

Library of Congress Card Catalog Number:
2008910530

cover image © Domen Colja | Dreamstime.com

Published by

American Literary Press
Five Star Special Edition

8019 Belair Road, Suite 10
Baltimore, Maryland 21236

Manufactured in the United States of America

A TRIBUTE

To Lancelot Law Whyte and Leo John Baranski,
founders of the Unitary Field Theory, the basis of this work

CONTENTS

Acknowledgments   xi

Introduction   xii

Chapter 1: The formation of stars, atoms and molecules

The physical universe   3
The birth of stars and the formation of atoms   6
Atomic organization and the process of nucleosynthesis   7
The formation of water and simple organic molecules   7

Chapter 2: The formation of the solar system

The emergence of the solar planetary system   11
The formation of the earth   13
Formation of the atmosphere and oceans   13
Formation of the crust   14
The life facilitating properties of earth   14

Chapter 3: Water as matrix of life

A cooperative structure wave in water   18
The concepts of water cluster, tension and pressure   19
An active role for water in protein structure   20
The intracellular gel and the local generation of motion   21
Water as promoter of communication and chemical change   22
Water as agent of retardation of physical velocities   23

Chapter 4: The universal process and time flow

A new conception of matter and time   26
Introduction to the unitary principle   27
The unitary field and its process   29
The measurement of time   30
Physical time vs. psychological time   31

Chapter 5: The cell as unit of life

The operation of the unitary field   36
The unitary properties of living protein   36
The property of chirality   37
Structure, process and complexity   39
Organic stability   40
The unitary view of evolution   43
Cell as biological unit   46

Colored illustrations   51

Chapter 6: The animal organism

Organisms vs. networks and mechanisms   61
Order from process   63
Cooperativity and self-regulation   67
Field asymmetry as life energy   69

Chapter 7: The human nervous system

The unitary view of brain organization and functioning   74
The thinking process and its states   80
The structural meaning and nature of consciousness   82
Historical development of the thinking process   82
The symbolic system of language   88

Chapter 8: The nature of physical reality

Cognitive aspects of reality   92
Perceptual aspects of reality   93
The fourth spatial dimension   103

Chapter 9: The socio-cultural environment

Sources of socio-cultural disorder   108
The universal process as source of order   111
Towards unitary order in society   112

Index   115

ACKNOWLEDGMENTS

I thank the librarian staff of Providence VA Medical Center, particularly to Ms. Cheryl Banick and my special friend Timothy Gormally, for their courtesy and help over the years. I also express my appreciation to my niece and good friend Maria Antónia for help with the computer and interest in my book. Thanks also go to my son Alex and my daughter-in-law Laura for their support and continued interest, and to my grandchildren Michael, William and Katherine for being sources of mental energy. Last, but far from least, I thank my wife and colleague Angie for her understanding, constant encouragement and support, without which this book could never have been written.

INTRODUCTION

Life as manifested to us is a function of the asymmetry of the universe and of the consequences of this fact.
Louis Pasteur, 1857

We live in a universe of forms in constant change. Local forms appear, develop, and disappear in a cyclic, ordered way. These forms did not emerge spontaneously but were formed from simpler ones by the same one-way process of development. This universal formative process has defied human understanding for millennia, yet it was scientifically described more than half a century ago. It appears, however, that this important achievement has escaped the attention of science.

The continuity in the sequence of change must ultimately connect biological to physical systems. This must be so for life to exist within the physical universe. Classical quantum field theory recognizes this fact by admitting that the long-range electromagnetic force is involved in the generation and maintenance of life. Accordingly, communication between space and organism and within organism is conceived to be effected by wave patterns. But this advanced physical theory is based on a reversible concept of time and cannot explain the mysterious one-way formative process.

To describe irreversible processes a one-way physics is required. This type of physics, a special quantum field theory, was formulated by the Scottish physicist Lancelot Law Whyte in his work entitled "The Unitary Principle in Physics and Biology," published for the first time in 1954. The unitary field of Whyte was later extended by Leo Baranski, an American psychologist and unitary theorist. The unitary field is considered to be made up of polarized three-dimensional quantum structures or particles in continuous alternating change. These structures are the precursors of all structural forms. In the unitary conception, the electromagnetic field becomes a polarization field, and waves become pulses of quantum field particles.

The new theory replaces the classical concepts of permanent matter and free energy by those of changing structure and structural asymmetry. Structure is taken to be a system of orderly spatiotemporal relations that, to fit the requirements of complex biological structure, are all asymmetrical. A specific structural pattern lies deep in each structure. The formative or structuring process is due to an intrinsic tendency in microstructure for structural asymmetry to decrease. This tendency is structural, not teleological, and is counteracted by the tendency of the field as a whole to restore asymmetry. In biological systems these two tendencies correspond to anabolism and catabolism.

Unitary theory is universal. It is applicable to all natural systems, simple or complex, physical or biological. It explains in a scientific way both our relations to the universe surrounding us and the creation of all forms in existence, including the forms of our thoughts, feelings and behavior. Mental processes, which puzzle current psychology, are not different from any other formative processes. They involve, however, the concerted action and multiple transformations of aggregates of quantum field particles supported by the nuclear framework of the organic atomic matrix.

Biological processes are far more complex, and therefore much slower, than physical processes, and for this reason we can better appreciate in them the universal process in operation and the irreversible character of time. Present physics is dominated by relativity theory, but this space-time theory is only valid for physical systems. Its concepts do not apply to biological systems, and an attempt to do so only leads to confusion between biological time and physical time. It is in this confusion that the "paradox," in reality impossibility, of time travel lies.

Unitary theory combines the minimum spatial and temporal relations that are required for any general theory of process and thus may express the true law of nature. It is more general than any other theory and is thus capable of unifying all the sciences. Furthermore, by clarifying the intimate relation between physical nature and human nature, it modifies our way of thinking and behaving, making us willing to cooperate with each other. This cooperation can form the basis of a new social order capable of unifying humanity.

In a previous work related to the physics of living processes, we briefly exposed the unitary principle together with a number of theories pertinent to specific aspects of the functioning of biosystems. In the present work, we use unitary field theory to describe the evolutionary chain of events from field level to social level. As a first step towards unification of knowledge, an attempt is made to interpret in unitary terms all the specific partial theories employed.

Unitary theory has given us the intellectual tools for the study of the enormously complex organic realm. A new sub-atomic biology, unitary biology, has been created, which considers the universe and the environment as agents of the living process. With the admittance of a field and knowledge of its formative process, the complexity of organic structure has become less intimidating. But until the theory can be of service to disciplines such as those of medicine and psychology, a considerable amount of work is required.

At present, formidable theoretical and technical obstacles are blocking the road to further development and application of this process theory, but they are not insurmountable. A bright future lies beyond them, but to reach it physics and biology must work together. As far as we know, the theory has not been tested, let alone confirmed, by physics and has therefore no authority, but the universality of its concepts and the trend of history will, sooner or later, bring unitary theory into prominence. Its scientific description may then be regarded as one of the greatest scientific achievements of the twentieth century.

For this to occur, unitary concepts must be transferred from the biological systems to the physical systems. Replacing mathematics by physics in the parody used by the mathematical physicist Stanislaw Ulam, we can express this need more forcefully: "Ask not what physics can do for biology, ask what biology can do for physics".

FROM STARS TO SOCIETIES

CHAPTER 1

THE FORMATION OF STARS, ATOMS AND MOLECULES

Protons have a special value in nuclear, atomic and chemical physics.
P.C.W. Davies, 1982

Modern physics theory regards every substance, body or system as ultimately built up of atoms. Whether free or bound into molecules or macromolecules, atoms constitute the common basic units of all things. Despite striking dissimilarities of form and quality, a rock, an apple, and ourselves are all related by way of their constituent atoms.

Atoms, in turn, are also related to one another. They form a close family, which originated from a basic atomic unit—the hydrogen atom. And the hydrogen atom is itself a product of the basic energy field of our universe. Thus, it is through the universal field that everything is ultimately related to everything else.

But atoms do not pile up arbitrarily. Every atomic-molecular structure, simple or complex, is differentiated into a specific form. This implies the operation of some underlying organizational process that brings about the individuality or uniqueness of things. In doing so, it develops the physical, chemical, and biological systems.

Those links to the universe make us a part of nature. We and nature are just one. But this special relationship acquires an even greater meaning and beauty when it is scientifically understood. On the way toward such understanding, we should consider first the processes whereby atoms and molecules were formed by past stars. This is equivalent to reviewing the history of the universe.

The physical universe

Over the last decade we have witnessed a remarkable progress in cosmology, derived mostly from data obtained by ever more potent and sophisticated telescopes, which now cover almost the whole electromagnetic spectrum. By the turn of the century, over 100 billion galaxies had been identified, each containing hundreds of millions of stars! These unexpected observations have led to insights into the nature and workings of the universe that are difficult to reconcile with present-day physical theories.

We must admit that our knowledge of the composition of the universe, at a fundamental level, is still very incomplete: the ordinary matter we know about, which makes up all the stars, planets and our bodies, accounts only for less than 5 percent of the total mass of the universe. The bulk of the material remains unidentified and is at present aptly named *dark matter* and *dark energy*. We are also uncertain as to the basic nature and relations of the physical forces. Perhaps our astrophysical level of understanding is still too meager to allow the formulation of a truly scientific theory of the physical universe.

The most prevalent theory on the subject, with roots in the mathematics of general relativity, is the so-called big-bang theory. It proposes that the universe originated from a very hot, very dense state in a big-bang explosion and has been expanding ever since. What triggered this event is not clear since there is no record of the processes that immediately preceded the bang, but the explosion has been attributed to the action of a swift period of inflation during which an initial speck of matter/energy grew exponentially, doubling its size every 10^{-37} seconds for a tiny fraction of a second. It was brought about by a *repulsive gravitational field*, which, according to particle physics, can be generated in extremely dense states of matter.[1] The energy left from the antigravitational field was released as photons, some in the form of gravitational waves; matter spread its content of particles evenly into the hot primordial space in the form of ionized gas, and the universe—and time—was on the move.

This interpretation of initial cosmic events has long been challenged by implications derived from powerful and logical field theories that favor a universe with no beginning and no end (and therefore with no inflation or expansion) and always in constant change. In this view, the universe self-constructs and self-destructs continuously, in cyclic fashion.[2-4]

From the overall distribution of atomic matter in the universe, we know that 70 percent comprises hydrogen, 27 percent helium and the remainder 3 percent includes all other atoms. Hydrogen atoms have a single proton as nucleus, surrounded by a single orbiting electron. The proton is a complex and extremely stable structure made up of smaller particles called quarks and gluons. According to the classical big bang theory, all existent hydrogen atoms, which are the constituents of stars, were formed during a short time interval after the brief and single explosion of matter/energy.

Given the astounding number of stars now known to exist, some of them tens of times more massive than our sun, it is inconceivable, as Paul Hollister recently pointed out,[5] that a single explosion could have produced so unimaginably enormous quantities of hydrogen. On logical grounds alone, some other major source or sources of hydrogen production must exist.

Astronomical data have consistently shown an extremely bright object located at the center of all large galaxies. Its diameter is only one millionth the size of the host galaxy but it radiates as much energy per second as a thousand galaxies. This extremely bright burning orb is known as a *quasar* (for "quasi-stellar"). Paul Hollister has therefore proposed that quasars are galactic geysers of hydrogen nuclei and electrons. Since the stars and gas clouds near quasars have enormous orbital velocity and the jets exploding outward can reach distances of thousands of light years, it is believed that the power source of this super massive density is a *black hole*. Precise measurements of the mass density of galaxy M87 showed it to be equivalent to three billion solar masses compressed into a space no larger than the solar system.[5]

The concept of black holes was derived from general relativity theory, which predicts that if matter is sufficiently compressed its gravity becomes so strong that it carves out a region of space from which no object, not even light, can escape. If an object falls in, it cannot get out.[6] Black holes are believed to be formed in the gravitational collapse of massive stars at the end of their lives. Paul Hollister now proposes that gravitational collapse of a gigantic nebula of dark matter could also have led to the formation of the super massive density at galactic centers.[5]

Events occurring in the basic field are not visible to us, but the theory assumes that the physical principles that form stars are also at work in the formation of super massive galactic densities, although at a vastly larger scale. The implosion of dark matter is attributed to gravity, but the negative pressure of dark energy and other properties of the basic field may possibly contribute.

Whatever the forces involved, as implosion progressed the vibrating quantum particles of dark matter became more concentrated, the number of their collisions increased and temperature started rising. Over eons of galactic time at temperatures of thousands of trillions degrees Kelvin (1 trillion $°K \cong 1$ trillion $°C \cong 1.8$ trillion $°F$), free quarks and gluons were assembled and an extremely dense and glowing quark-gluon plasma slowly emerged. Further gravitational compression of the ever-denser nebula suddenly collapsed a portion of the core, raising the temperature perhaps over 10^{30} degrees Kelvin. With the threshold of *quark-gluon fusion* reached, stable protons were formed by the attractive action of the strong force, and with them charge neutralizing electrons also appeared. Proton-electron plasma of much lower density was thus generated. When thermonuclear fusion energy exceeded gravitational energy, the lighter proton-electron plasma was explosively ejected into the freezing surrounding space. Upon cooling, clouds of atomic hydrogen were formed, which organized into the vast number of stars that constitute a galaxy. Over billions of years, when the above forces finally reached equilibrium and the explosive process came to an end, new stars could no longer be formed, and the old ones, after exhausting their fuel, ended up their lives collapsing into black holes.

According to this theory, each galaxy has its own source of hydrogen and therefore should be viewed as a unit in the universal organization, as a cell is a unit in the biological organization. But structural organization is a self-limiting process. Stars are born and die, and galaxies appear and disappear. Under the appropriate circumstances, these events occur at any time and in any place in space. In the present cosmological era, billions of galaxies are glowing across the visible universe, distributed throughout in a rather homogeneous fashion.

Instead of sucking in matter, as classical black holes do, the super massive black hole at center of galaxies ejects it. It is a black hole working in reverse. When its action is combined with that of the much smaller but numerous black holes resulting from the collapse of old massive stars located at the periphery of galaxies, a cycle of stellar mass-energy is established, going from basic field to stars to basic field again. Seen in this light, each galaxy is no more than a temporarily organized circulation of energy.

The science of modern cosmology rests on the equations of general relativity, which describe the global behavior of matter and energy and of space and time. The universe is assumed to have evolved from initial conditions imprinted in the first microsecond after the big bang and to be locked to a single line of evolutionary time. If Hollister's theory is confirmed, the above concepts must be changed and alternative explanations found for a number of phenomena that appear to support the big bang theory.

In a broader view of things, however, Hollister's theory may just be one aspect of a more fundamental creative-formative process that governs the universe. Since the operation of this process, known as the unitary principle, is more apparent in complex structures, such as living systems, its description will be deferred to later chapters.

The birth of stars and the formation of atoms

Gravitational implosion-energy explosion is also the basic process of star formation, both in the classical big bang theory and in Hollister's theory. This type of creative process appears to be the one chosen by nature. The above theories, however, differ in their concepts of space, time and hydrogen source, and so the evolution of galaxies and age of their constituent stars are interpreted differently.

Our interest in this subject does not reside primarily in cosmology but in the formation of those atoms that later led to a new realm in the universe—the organic realm. Atoms, however, are born in the interior of stars, and it is the intrastellar processes of atomic assembly that are emphasized here. The Hollister's theory will be adopted to describe the galactic events that presumably led to star formation.

From the jets of hydrogen plasma propelled into regional space by the super massive black hole engine of quasars, clouds of molecular hydrogen were formed. These clouds rotated at high speeds, but over long stretches of galactic time the compressive gravitational force overcame the dispersive centrifugal force generated by the rotation. As a result, the clouds organized themselves into dense nebulae of ionized hydrogen, the mothers of future stars.

As the implosion process went on, the nebulae became denser and denser. Collisions among particles increased and core temperatures rose. Further compaction and higher densities eventually led to sudden collapse of a portion of the core, thrusting the temperature to millions of degrees Kelvin. When the threshold of nuclear *fusion of hydrogen into helium* was reached, luminous stars made their appearance. It may take thousands of years for stars to burn away their mother nebulae and to reach their full splendor when observed. This process is universal and should also have occurred in our galaxy, the Milky Way, so-called because ancient Mediterranean civilizations believed it to be composed of milk spilled from the breast of a goddess.

The stars of the first generation were massive bodies, tens of times bigger than our sun. They burned hydrogen so fast that most of it was exhausted after a few million years. As the nuclear fusion process went on, gravitational forces were overcoming internal repulsive forces, induced by accumulated electrons, and forcing the partially collapsed core to contract even more. With core size diminishing, temperatures in some central regions went up so high that helium nuclei started burning. At over two billion degrees, and after a succession of reactions, three helium nuclei fused— and the *carbon nucleus* was formed. And when a fourth helium nucleus was fused with one of carbon, the *oxygen atom* was generated.[7]

In the most massive stars, and at still higher temperatures, complicated reactions involving pairs of carbon and oxygen nuclei yielded such elements as nitrogen, sodium, magnesium, silicon, phosphorus and sulfur. With continuing gravitational implosion, core temperatures reached several billion degrees and some silicon nuclei broke down into helium nuclei, which added to other silicon nuclei produced chromium, manganese and iron.

The fusion of iron did not release, and even required, energy. When this stage was reached, the stars very quickly ran out of useful fuel, and the cores, under intense pressure, caused the stars to explode. The extreme pressures and temperatures within such explosions are thought to be sufficient for the formation of all other nuclei heavier than iron.[8,9] Ultimately, the peripheral remnants of stars were ejected far into interstellar space to become cosmic dust—and the seeds of future stars and planets—and the cores eventually collapsed into black holes.

Atomic organization and the process of nucleosynthesis

The building up of atomic structure represents the first level of energy organization in the physical world. Its basic unit is the hydrogen atom. The formation of heavier nuclei from lighter ones is the end result of a balancing among the forces of nature, most particularly the nuclear forces (weak and strong) and the electromagnetic force, occurring in a high-energy environment. Each nucleus synthesized is a compromised structure of certain stability, possessing a given energy level. For energy to organize this way some time is required.

Stars contain ordinary hydrogen (one proton, one electron) as basic fuel. Protons are very complex and unique structures. They cannot be fused into helium-2 (two protons), which does not exist in nature. Even the *strong nuclear force,* which is active in a thermonuclear bomb, is not strong enough to fuse them. If it were, all the hydrogen in the universe would soon be exhausted and we would not be here. The helium that exists in stars is helium-4 (two protons, two neutrons), the neutrons being formed by the fusion of electrons with protons. The force leading from hydrogen nuclei to helium-4 nuclei is the so-called *weak nuclear force,* which is 10^{18} times slower in action than the strong force, at the same density and temperature.[10] The time delay provided by the weak force may have facilitated the organization of protons and neutrons into atomic nuclei.

For the process of nucleosynthesis to be self-sustainable, earlier nuclei should always be available and produced in amounts high enough to maintain the rate of synthesis of later nuclei. Crucial to this process is the positioning of the energy levels of the nuclei involved in a reaction, which is established by complicated interactions between nuclear and electromagnetic forces. As it happened, the energy level of three helium-4 nuclei plus the thermal energy level of the star interior matched almost exactly the nuclear energy level of carbon. A *resonance* was thus set out, which led to the production of huge amounts of carbon, and consequently of oxygen.[11]

Fortunately, the nuclear energy level of carbon plus helium-4, let alone the added thermal contribution of the surroundings, is higher than that of oxygen and so no oxygen resonance is possible. In this way, the formed carbon was not exhausted and was used later advantageously for the energetics of life.

The formation of water and simple organic molecules

Contrary to the formation of nuclei, which requires extremely high temperatures to be accomplished, the formation of molecules can only go on under colder temperatures. The colder the temperature is, the higher can be the molecular complexity. At temperatures less than 3,000 °K, which are reached in the outer envelopes of some stars during the so-called *'red giant'* phase of evolution, CO (carbon monoxide) and H_2O (water) molecules are already forming and leaving the star much before its final collapse.

This phase occurs during helium burning, after exhaustion of most of the hydrogen. As the core contracts, the outer shell of the star expands like a balloon. The expansion weakens the gravitational attraction to the distant core and the outward radiative flow coming from the center is then capable of gently expelling the molecular material from the outer layers of the star into the surrounding interstellar space.[7]

It was, however, within the interstellar medium that most of the primitive molecules were formed. How the process was accomplished in a medium with an average density of less than one atom per cubic centimeter is less clear. Presumably, it occurred in isolated condensations of hot gas and dust close to the stars and in farther away, denser, and much colder clouds where the temperature could reach as low as 20°K (-253°C). Spectroscopy and radio astronomy confirm the existence in these clouds of hydrogen, water, several carbon and nitrogen compounds (methane, ammonia) and other more complex organic molecules made up of a dozen of atoms or so (alcohols, ethers).

Solid grains of silicates and other materials, enveloped in a shell of water ice, have also been found in the interior of these clouds. How they come to form there is not known, but once formed they would effectively cool the cloud and made possible the assembly of more complex molecular forms.[7]

SELECTED REFERENCES

1. GUTH A. The inflationary universe. Perseus Press, Oxford (1998).
2. BARANSKI L. Scientific basis for world civilization. Unitary field theory. The Christopher Publishing House, Boston, USA (1960).
3. AXELSSON S. Cholinesterases and unified neurophysics. NeuroQuantology 3, 59-71, 2005.
4. YOUNG RR. The steady-state galaxy theory (2005). Complete manuscript online at *http://www.galaxytheory.com.*
5. HOLLISTER P. The origin and evolution of the universe, a unified scientific theory (2004). Complete manuscript online at *http://www.origin-of-universe.com.*
6. CARR BJ, GIDDINGS SB. Quantum black holes. Scient. Amer. 292, 48-55, 2005.
7. KANDEL R. Water from heaven. Columbia University Press, New York (2003).
8. COWEN R. Galaxy hunters. The search for cosmic dawn. National Geographic, 203, 2-29, 2003.
9. HENRY PP, BRIEL UG, BOHRINGER H. The evolution of galaxy clusters. Scient. Amer. 279, 52-57, 1998.
10. DYSON FJ. Energy in the universe. Scient. Amer. 225, 51-59, 1971.
11. BARROW JD. The constants of nature. Random House, London (2002).

CHAPTER 2

THE FORMATION OF THE SOLAR SYSTEM

Water and energy always flow and change their form on our planet,
but they never vanish, nor are they created.
Robert Kandel, 2003

Biological life requires a changing environment. The changes allowed must be not too great or too abrupt, so living systems can adapt to them. Otherwise life would not develop, or if already developed, it would die out. For human life, others requirements must be met. It needs, among other things, a solid foundation, a liquid medium (water in our case), and a source of energy.

From those requirements, a star can only supply the energy. Its high temperature, gas medium and tumultuous internal and external change are incompatible with the development or maintenance of life. Only a solid planet, located at an appropriate distance from a star, possessing abundant water, and whose environment is relatively stable can give rise to life.

In our universe, these requirements were met by the planet earth, belonging to the solar system, not immediately after its formation but after a period of about three billion years. During this interval, the earth grew up to its full size and suffered tremendous physical and chemical transformations that ultimately led to the creation of the oceans, the continental crust, and an appropriate atmosphere.

If the physical conditions are appropriate, life may exist or can develop on any of the forty-seven extrasolar planetary systems already identified or on the billions more thought to exist throughout the whole universe.

The emergence of the solar planetary system

Our Milky Way galaxy was formed fifteen billion years ago. For more than eight billion years thereafter there was no sun or earth but only massive stars and dust clouds producing ever more atoms and simple molecules, most notably water. The intense ultraviolet radiation emanating from the massive stars carved out ionized cavities in the dense molecular clouds.

In a region located about 25,000 light-years away from the galaxy core, compression by an advancing ionizing front from a neighboring star led a major cloud of hydrogen gas and dust to initiate gravitational contraction and spinning. Under the competing forces of gravity, gas pressure and increasing rotation, the shrunken cloud became a low-mass star, and the circumstellar disk, partially eroded by photoevaporation, became the protoplanetary disk.[1,2] When the neighboring massive star finally exploded, the low-mass star emerged as our sun and a few gravitational condensations around special points in the unstable disk later became the embryonic planets. These events occurred about 4.5 billion years ago.

In the universe everything is spinning, including the clouds of matter that form stars. While clouds contract by the mutual gravitational attraction of their matter particles, becoming denser and hotter in the process, their spin rates must concomitantly increase. This occurrence results from the law of conservation of *angular momentum*.

Angular momentum is a property of an object or system of objects undergoing circular motion in a plane about a point. An increase in either the mass, the velocity of a rotating object, or the distance the mass is situated from the center of rotation causes an increase in angular momentum and vice-versa. Being a conserved property, the total magnitude of angular momentum remains unchanged even though there may be exchanges between components of a system. In the case of a rotating cloud, as the mass decreases the rotating speed must increase.

In the case of the solar system, in spite of the planets containing only 0.13 percent of the mass of the sun, their orbital motions account for nearly all the orbital angular momentum. The high orbital velocities of the planets in relation to the very slow spin of the sun on its axis (twenty-four to twenty-seven days) and the much longer distances of the planets to the sun in relation to the sun's radius came to overcome by far the solar mass effect on the overall angular momentum of the system.

Therefore, during the evolution of the solar system cloud, a transfer of angular momentum from its swirling center, the protosun, to the surrounding protoplanetary condensations must have occurred. It is believed that this transfer process depended on the existence of magnetic lines of force that accelerated the distant cloud's rotation at the expense of the spin of the central condensation.[1]

The sun is a typical star in the hydrogen to helium burning phase. It has a diameter of about 1,392,000 km and is 149,597,870 km away from earth. This distance corresponds to 499 light seconds, so sunlight takes about eight minutes to reach earth. Solar radiation is produced in the inner core of the sun and then transferred to the surface by radiation and convection processes, which take several hundred years to accomplish. As a result, the radiation we receive on earth every day was created in the sun a very long time ago.[3]

The solar system, that is, the planets and the sun itself, rotates from left to right, in a counterclockwise direction. This directionality may have influenced the organization of matter/energy during evolution of life on earth. But that aside, the system would work equally well if it were rotating in the opposite direction. Time cannot be measured by a system that can go forward and backward, so the solar system is time indifferent. Systems that behave this way are called *reversible systems*. To be reversible, systems must be stable. And stability implies no change in their trajectories or only slight changes in a predictable way. It was from the study of the solar system trajectories that Newton derived the laws of motion.[4]

The formation of the earth

The earth, and all the rocky planets, was formed by accretion, that is, by stepwise aggregation of cosmic gas and dust into its embryo in the solar cloud. As the proto-earth diameter increased, its gravitational force became strong enough to attract nearby clumps of dust grains the size of tiny planets (planetesimals), which contained condensates of oxides, metals and silicates. Earth bombardment by planetesimals or meteorites, some of them perhaps the size of the moon, went on for millions of years.

Data from tungsten isotope composition of meteorites indicates that most of the planet earth was formed about ten million years after the formation of the solar system. Accretion was effectively complete at thirty million years, when a mars-sized impactor hit the earth off center, blowing out rocky debris. A fraction of that debris went into orbit around the earth and aggregated into the moon.[5]

On colliding with the surface, these large bodies of matter produced immense heat. The extremely high temperature melted much of the earth, prompting the denser molten metals, mostly iron, to trickle down between the solid silicate mineral grains to the center of the planet where a magma ocean was formed. From this separation, the central core and mantle later evolved.

Situated some 200-400 kilometers underground, the magma ocean was under very high pressure and temperature, the latter sustained by further collisions and by radioactive decay. It remained active for millions of years, giving rise to volcanic eruptions and lava flows, which later led to the formation of the atmosphere, the oceans, and ultimately the continental crust. At this stage no life was possible.[6]

Formation of the atmosphere and oceans

All water existent on earth arrived from space during the accretion phase. At first it arrived within the solid or frozen granules that formed its embryo, and later within the meteorites, comets and larger bodies that constantly slammed its surface over the first thirty million years. With the separation of the metal core, this water was confined to the outer layers of the earth.

When the volcanic eruptions started, huge amounts of water in the form of vapor, together with volatile substances— largely carbon dioxide, methane, ammonia, and hydrogen sulfide, but very little oxygen— were projected into the atmosphere. This process, called "outgassing" of the mantle, went on for about fifty million years after the formation of the moon. Meanwhile, the atmosphere became hot and dense, saturated with water vapor which extended deep into space.

When the earth's surface started to cool off, the critical temperature of water (374°C) in the atmosphere was subsequently reached. Above this temperature, no amount of pressure could form water from the gaseous vapor. With the temperature now decreasing, the tremendous pressures then existent could finally cause water to rain. Hot at the start, this rain began far out in space and subsequently deluged the earth, not for forty days and forty nights but perhaps for thousands of years, and in the process the primary oceans were formed.[1] And when it was all over, the sunlight shone through.

The oceans now cover 71 percent of the earth surface and their volume has hardly changed for the last two billion years. With an average depth of 3,700 meters, they contain 1.35 billion km^3 of water. This may sound too much but it is just the equivalent of 0.02 percent of the total mass of the earth. As far as it is known, no other planet in the universe contains such a vast amount of water.

During those times, sunlight was much fainter than it is today. The surface of the earth would freeze over if not for the greenhouse effect. It was caused by the presence in the atmosphere of heat-absorbing molecules made up of three or more atoms, most particularly CO_2. These molecules prevented the thermal infrared radiation reflected from the earth's surface from escaping into outer space. In this way, they kept the lower atmosphere and the surface of the earth relatively warm.[1]

Formation of the crust

Going from the center of the earth upwards we encounter three different layers: core, mantle and crust. The thickness of the crust varies from about 8 km under the oceans to about 40 km under the continents. Our understanding of the processes that led to crust formation is far from complete. Presumably, the crust was formed by a process akin to that of volcanic eruptions, which originated not from the deep hot core but from the somewhat milder upper mantle and lower crust.

At these depths, the temperature is close to 900°C and therefore high enough to melt silica and silicate minerals. Silica melt is of low viscosity and could easily ascend, through fractures and cracks to the surface, where it spread in large horizontal sheets in the upper crust. These sheets then cooled and crystallized to form granite rock, which makes up 70 to 80 percent of all continents, the rest being basalt, a rock high in magnesium and calcium oxides. A similar process took place in the formation of the ocean basins, which are made up almost totally of basalt.

Perhaps the most significant feature of the composition of the earth's crust is that it is dominated by comparatively few elements. In addition to *oxygen* and *silicon,* which are the most abundant, only six more elements—*aluminum, iron, calcium, magnesium, sodium, and potassium*—are present in amounts greater than 1 percent.

The earth's crust is not a single stable shell. It is divided into huge plates, 80-400 km thick, which drift atop the soft, underlying mantle. The first continents were formed 400 millions years after earth's formation. Since then, they have drifted all over the globe until they reached the positions and shapes they have today. They continue to move even now: Europe and North America are drifting apart 3 cm every year, and the same is true for Africa and South America.

Until now we have outlined, in broad brush strokes, the continuing major events that occurred in our tiny corner of the universe, from the birth of our galaxy until the earth reached a stage of relative quietness. In the human time scale, we are approximately 3.9 billion years ago.

The life facilitating properties of earth

In the chemistry of its solid mass, oceans and atmosphere, the earth possessed all the ingredients necessary for the building up of a new kind of structure—biological or living structure—based not on the organization of atoms, as is the case for physical structure, but on the organization of molecules and macromolecules. The higher physical level supporting biological structure continues to be based on, and controlled by, the atomic level below. However, its information processing power is vastly greater than that of the atomic-based physical structure. This allowed it to reach degrees of complexity which could never be attainable in the physical realm.

Moreover, the position of the earth in the solar system, the presence and size of its moon satellite, and a few intrinsic features appear to have endowed our planet with life facilitating properties. In effect, its distance from the sun allows it to receive enough heat energy to maintain surface water in the liquid state; its daily spin allows the daytime side to warm in the sunshine while the nighttime is cooling; and the mild tilt (23° 27') of its axis of rotation relative to its axis of motion around the sun gives it the seasons and tides that moderate our climate. The latter effect is a consequence of the gravitational pull of the moon.

The earth itself, by its size and density, possesses a gravitational attraction high enough to hold a wide variety of light molecules, such as water, in the atmosphere, thus preventing them from drifting into space, and low enough to allow small drops of rain to fall. If the force of gravity were not as small as it is, the water cycle would not be possible. We would have no clouds, no rain, and therefore no life. Furthermore, magnetic forces emanating from its iron core—the earth's magnetic field—deflect back to space the solar wind, a continuous flow of highly energetic radiation (mostly protons and electrons), which otherwise would destroy life if free to bombard the earth's surface directly.

During this period, and in spite of all these life-protecting and facilitating properties, the earth's continental crust was still a very hostile and barren environment. The soil was poor in nutrients and the lack of shade made it vulnerable to desiccation by constant exposure to the bright light of day. Furthermore, there was very little oxygen in the atmosphere, and therefore practically no ozone, which is a good filter of solar ultraviolet radiation. The photons carrying this type of radiation have enough energy to disorganize the macromolecular building blocks of living structure, that is, proteins and DNA. Life on dry land did not appear to be feasible at this stage.

SELECTED REFERENCES

1. KANDEL R. Water from heavens. Columbia Univ. Press, New York. (2003).
2. HESTER JJ, DESCH SJ, HEALY KR, LESHIN LA. The cradle of the solar system. Science 304, 1116-1117, 2004.
3. AKIOKA M. The sun and its activity. Biomed. Pharmacother. 56, 243s-246s, 2002.
4. ROTHMAN T. Irreversible differences. The Sciences 37, 26-31, 1997.
5. JACOBSEN SB. How old is planet earth? Science 300, 1513-1514, 2003.
6. ALLÈGRE CJ, SCHNEIDER SH. The evolution of the earth. Scient. Amer. 271, 66-75, 1994.

CHAPTER 3

WATER AS MATRIX OF LIFE

*There seems to be no simple molecule that can mimic
all of the useful biological functions of water.*
Philip Ball, 2004

There is general agreement that life started when liquid water became available. What is less certain is where the assembly process was first initiated. According to recent views, life was developed in porous mounds of freshly precipitated clay on the bottom of the oceans.

These mounds are formed around hydrothermal vents (springs), ejecting diluted molten rock from the upper mantle to the ocean crust. This rocky magma, rich in chemically reduced minerals, such as sulfides, iron, magnesium and other metals, dissolves in the nearby water, raising its temperature from 2° to 300-400°C.

It is believed that the first unicellular organisms (bacteria) were created in this infernal environment. Without access to sunlight, they used the energy that emerged from the earth's interior in the form of heat and in the form of electrons resulting from the biochemical oxidation of the reduced minerals–*chemosynthesis*.

A new type of bacteria, called cyanobacteria, then evolved, migrated to the surface and developed the capacity of using solar radiation as source of energy. This process—*photosynthesis*—uses the energy of sunlight to break the molecules of H_2O and dissolved atmospheric CO_2 to form carbohydrates (COH) and molecular oxygen (O_2). Then the primitive plants, known as blue-green algae, incorporated cyanobacteria as sources of their own energy. Over time, the oxygen produced photosynthetically by green algae increased the concentration of oxygen in the atmosphere.[1]

It took approximately two billion years for atmospheric O_2 to reach levels comparable to those existent today. Meanwhile, when enough O_2 became available, bouts of O_3 (ozone) were being formed whenever lightening occurred and in a smoother way by oxygen collision with ultraviolet sunrays. Around 2.5 billion years ago a protective stratospheric ozone layer was finally in place around the earth. Only then could life move safely from sea to dry land, and the slow and tortuous process of surface evolution could get underway.

Today more than 62 percent of the biosphere is still living in the sea, either at the ocean bottom at 2° or around hydrothermal vents at over 100°C and at depths over 1,000 meters, which correspond to a hydrostatic pressure higher than 100 atmospheres. Only less than 40 percent of the mass of living organisms inhabits the surface of the earth.[2]

We have given a very sketchy view of events that preceded the appearance of life on this planet. Because of its requirements, life could not have developed before it did. For life to appear, millions of stars had to be born and die, water had to be present in sufficient amounts on the surface of the earth and, later, a protective atmosphere had to be created.

This book is concerned mainly with the physical principles by which life operates. Since life has an aqueous basis, we should try to discover whether water possesses any special properties in its structure which allows it to serve so prominent a role.

A cooperative structure wave in water

The water molecule is a very dynamic structure containing one highly energetic oxygen atom partially tamed by two hydrogen atoms attached to it. The molecule has four fractional charges, two positive (corresponding to the protons of the hydrogens) and two negative (corresponding to the electrons of the hydrogens attached to the oxygen). It is thus electrically neutral overall, but its charge distribution is asymmetrical. The oxygen atom is more electronegative than the hydrogen atom, and the shared electrons are forced to dispend more time closer to the oxygen than to the hydrogen. This charge asymmetry creates a permanent electric dipole of very high moment.

In three-dimensional space, the charges are directed towards the corners of a near-regular tetrahedron or triangular pyramid. By mutually binding hydrogen protons to oxygen electrons, through so-called hydrogen bonds, the tetrahedral molecules of water can form among them three-dimensional networks.

Hydrogen bonds are ten times weaker than covalent bonds but more powerful than the nonspecific bonds that attract uncharged atoms to each other. This intermediate strength of the hydrogen bonds endows water molecules with tremendous flexibility. They can bend, stretch, break and link again and so water can easily adjust itself to the disruptive presence of any surface it encounters—proteins, lipids, carbohydrates or small solute. Water owes its living properties to the characteristics of its hydrogen bonds.

In terms of hydrogen bonding, the O-H group is unusual because it can act simultaneously as donor and acceptor of electrons. Furthermore, when involved in hydrogen bonding, the group becomes polarized in the surrounding electric field; that is, it develops an additional induced dipole. As a result, its donor and acceptor properties become enhanced because the oxygen is now more negatively charged and the hydrogen more positively charged. Under these circumstances, the contribution of the numerous O-H groups of an aqueous system to its overall stability is not only additive but cooperative.[3]

Cooperative phenomena result from interactions among the constituents of a system. In the case of water, these interactions are long-range, have a definite velocity, and may also have a direction. As a result of cooperativity, therefore, the oscillations created by the constant making and breaking of hydrogen bonds are propagated through the liquid as waves. There is a wave structure hidden in water. The build-up and break-down of the structured regions should appear as integration and disintegration of aggregates (clusters) of temporarily bonded water molecules. As cooperativity in water is based on induced polarizability, it links structure to energy.[4]

The concepts of water cluster, tension and pressure

The state of a substance—gas, liquid or solid—depends on the balance between the kinetic energy of its individual molecules and the intermolecular forces attracting them. Situated between the gas and solid states, the liquid state must be a peculiar one. Although constantly under pressure, liquids adopt a condensed shape as though they are held together internally like solids. It is the dynamic nature of their molecular structure which allows them to do so. They possess the properties of both gases and solids but in different degrees.

For the water network to maintain its flexibility, it must develop tension to oppose the disruptive molecular action of pressure. Pressure and tension must somehow coexist in water with equal force but without canceling each other out. Watterson has proposed that this occurs at separate physical levels: *pressure* on the macroscopic and *tension* on the microscopic level.[5]

From the gas law, he calculated that the volume occupied by one gas molecule at normal pressure and temperature is 40 cubic nanometers, which corresponds to a cube of approximately 3.4 nm sides. He reasoned that at the molecular level, and under equilibrium conditions, pressure must have the same meaning everywhere; he then transferred that volume to liquids and assumed it represented the smallest volume where pressure can exist. Watterson called it the *pressure pixel,* in analogy to the pixel concept of information science.

For water, the size of 3.4 nm corresponds to the length of 11 water molecules and the volume of 40 nm^3 corresponds to the packing of approximately 1,400 water molecules (Fig. 1, pg. 51). In the wave model of water, that is the volume of a cluster that is made up of that many water molecules bonded together by an inward-directed force derived from concerted hydrogen bond interactions. A water cluster is conceived as a very dynamic *water particle* which aggregates and disaggregates as a unit at time intervals of the order of 10^{-12} seconds.

The structure of water, then, is visualized as made up of sinusoidal cluster waves of 3.4 nm wavelength, in incessant polymerization and depolymerization. When the wave propagates, the single water molecules which momentarily make up a cluster remain stationary. What moves and propagates as waves are the clusters. In energy terms, the long-range clusters are carriers of structural energy in the form of vibrations of their hydrogen bonds. They also are the agents of action in all dynamic functions of water, and so they can be considered to be *water quanta.*

In classical statistical mechanics the agent of pressure is the single water molecule at microscopic level. There is no order at this level, just random motion of the molecules induced by thermal energy. It corresponds to the heat bath of current biology. In the model of Watterson, the agent of pressure is the water cluster. Below cluster level—that is, below the pressure pixel—pressure has no meaning. What exists is tension which opposes to, and equilibrates with, the pressure. As clusters form and break continuously, there is an on going alternation of order and disorder in water.[4]

Water clusters are very dynamic entities always adjusting themselves to the constantly changing surrounding conditions. Being dependent on water-water interactions, they are very sensitive to perturbations of these interactions by, for instance, the presence of solute. Introduction of solute into water restricts the movement of water molecules, particularly of those immediately surrounding the solute, the first hydration layer. It corresponds to a compression exerted on the liquid, and thus to increased pressure.

An increase in pressure must be met by a correspondent increase in tension. The water clusters therefore decrease in size and, to compensate for lost volume, increase in number. The decrease in size is brought about by the spring-like forces that restrain the lengths of the bonds and act as if the whole cluster were a single huge molecule. These forces behave like springs in that the more an outside pressure tends to distort the bonds, the more resistance it encounters. On the other hand, the increase in cluster concentration matches the concentration of solute molecules (Fig. 2, pg. 52). In terms of the wave structure, the solute-induced perturbation of reduced size and augmented tension of clusters is met by waves of shorter wavelength and higher amplitude.

One type of pressure most important in living systems is osmotic pressure. The force involved in osmosis is generated at the interface of a semipermeable membrane, such as the cell membrane, separating solutions of differing solute concentrations to which the membrane is not freely permeable. Here too, the water clusters, which are the agents of action in Watterson's theory, are the ones involved, and not single water molecules as assumed by current theories. A detailed account of the molecular mechanism of osmosis cannot be given here but can be found elsewhere.[6,7]

Life at the high hydrostatic pressures found at the bottom of the sea would not have been possible without the tensional equilibration mechanism offered by the special structure of water. Besides equilibration of pressure, water clusters are also instrumental in shaping the structure and function of proteins, the molecular motors of life.

An active role for water in protein structure

We have just described one of many cluster models of water structure. In the absence of a universal model, we singled out that of Watterson because it appears to us to be more fundamental than any other we know. Clusters as constituents of the liquid phase have been postulated to occur, and small clusters have been detected in water.[8] But the smallness of the timescales involved and the complex dynamics of hydrogen bonding have set size limits to what present experimental techniques can reveal and existent calculation methods for computer simulation can do. As it stands, studies have been limited to cluster sizes much smaller than those postulated by Watterson on physical principles. Confirmation of Watterson's theory is just not possible at present.

One major advantage of Watterson's cluster model is its adaptation to an underlying wave structure postulated for water and all other liquids. This accords well with the concepts of quantum mechanics and is also in line with modern quantum field theories of space-living matter energy exchange.[9,10] Furthermore, the model rests on the assumption that there is a fundamental size relation among gases, water and proteins, which appears to be confirmed by experiment. All known proteins are made up of one or more regions or domains, each possessing about 220 amino acids. In the functional state, the domains fold themselves into potato-like three dimensional structures or clusters which have been found to be similar in size to water clusters. Spatially, therefore, protein domains and water clusters fit together like hand and glove, giving rise to integrated, large-scale, subcellular structures. This is the architecture of living matter.

The water clusters adjust their size and shape to fit the specific geometry of domains and other solutes. Along the cell cytoskeleton, for instance, a one-cluster-thick water layer is formed, parallel to its protein filaments. The same occurs at the level of the lipid membrane. It is this packing of water clusters to protein domains and other solutes, brought about by tensional forces, that constitutes the *cytoplasmic gel*. The filament-associated water clusters are stationary, and therefore no solvent flow is possible, and this is the hallmark of a gel. This water-based gelation physics differs from the solute-based of classical theories, which requires a critical concentration of polymeric solute to achieve cross-linking and impairment of solvent flow.[11]

The stationary clusters form a stationary or standing wave. The corner of each cluster unit, where clusters meet, constitutes a *node*, that is, a region of minimal disturbance in the wave motion. Although the clusters are stationary, their individual water molecules continue to be active as if the clusters were propagating.

In Watterson's theory, the formation of stationary water clusters at solute surfaces, the so-called hydration layer, results from effects on water structure induced by the presence of a boundary. Since hydration layers are formed at any boundary, they must be a property of water rather than solute. They are maintained in place by lateral, inward tensional forces spreading from the water cluster to the inside of the solute cluster. Water cluster and protein domain, therefore, form a single system, structurally and functionally. This is a very important concept in the new biology.

The intracellular gel and the local generation of motion

The Watterson's concept of tension opposing pressure at mesoscopic level is unique and new in physics. Pressure is equilibrated by tension at the level of water clusters, whether these are free in bulk water or gelled to protein domains and other solutes at interfaces. Modern physics, deprived of the concept of tension, postulates that pressure is equilibrated by changes in the size of aggregates of water molecules at interfaces.

These aggregates, which correspond to the clusters in Watterson's theory, are structurally different from bulk water. They are the result of local compromises in water density, achieved by changing the strength of their intermolecular hydrogen bonds, in order to adjust their chemical potential, disrupted by local forces, to that of the whole water network.

The forces involved are mainly electrostatic and brought about by the presence of charged (hydrophilic) and uncharged (hydrophobic) regions at the surface of proteins. The charges, which are fixed and mostly negative, attract dissolved mobile counterions, which are thus distributed asymmetrically around proteins. The solutes accumulate in front of the charged protein patches and tend to decrease the activity, and thus the chemical potential, of resident water and to increase local pressure. These changes lead to the breaking of some intermolecular hydrogen bonds followed by compaction of the whole aggregate of water molecules by the surrounding pressure (high density water). In the process, the aggregate's chemical potential is raised to that of bulk water.

The water molecules adjacent to hydrophobic patches face a different dilemma: they cannot make bonds easily with the surface. Thus, they find themselves in a state of high energy and high chemical potential. This challenge is met by making bonds with themselves and stretching them with the available excess energy. The water therefore expands, decreasing its density (low density water) and lowering its chemical potential to bulk levels.[12]

Although the density changes of interfacial waters are just of the order of a few percent, they have profound effects on solubility and consequently on ion distribution. Furthermore, they affect the folding and unfolding of proteins directly and are thought to be at the basis of all biological motion. Only by working in concert with local water and ions within nanometer spaces can proteins be the nanoengines of life.

Interfacial or vicinal water is not only sensitive to pressure but to temperature as well. As temperature is increased, low-density intracellular water changes its properties not smoothly but abruptly, in jumps occurring at 15, 30, 45, and 60°C, suggesting it exists in four different phases. It is believed that the equatorial-polar temperature range on the surface of the early earth, together with requirements for a steady internal environment, set the human thermostat to the high level of 37.5°C, midway between the 30 and 45°C transitions. The other options were energetically less favorable.[13]

Under normal present life conditions, however, temperature and pressure remain fairly constant in the body. What constantly moves about are ions, and the electric and magnetic fields they carry influence the dipole moment, and possibly the structure, of protein-associated water molecules, which in turn influences, and is influenced by, the dipole moments and conformation of proteins. Water, ions and proteins form a single and inseparable functional system.

Low-density, ice-like water is very rigid and tends to maintain adjacent protein in an extended conformation. On the contrary, high-density, vapor-like water, with most of its bonds bent or even broken, does not hinder protein folding. Therefore, for protein to fold or contract, low-density water must be destructured by changes in surrounding ions, local pH, protein charge or any mechanical or electrical stimuli[14] (Fig. 3, pg. 53). A delicate balance of competing forces between water and protein must exist to allow for these swift, low-threshold changes to occur.

Muscle contraction, material transport, photon emission, cell division, secretion, pump function and many other functions all use *protein-gel phase transitions* as basic mechanism. In effect, phase transitions appear to be the common denominator of biological motion.[15] Coherent water-protein dynamics is thus the driving force of living systems. If this dynamics is impaired and the impairment is sustained, protein folding does not occur and pathologic conditions are sure to follow.

Hydrogen bonding and unbonding are at the basis of the transitions. Since they are cooperative processes, phase transitions once initiated always go on to completion. Their time scales, however, are longer than those for making and breaking hydrogen bonds in bulk liquid water, which are of the order of a few femtoseconds.

Water as promoter of communication and chemical change

The structure of bulk water can be seen as made up of two microdomains, low density and high density, in rapid exchange. The water network is continuous throughout the body. It extends from the intracellular to the interstitial space by way of channels in cell membranes and from the interstitial to the intravascular space by way of even bigger gaps in endothelium. There is thus a sea of water in the body. Within cells the water is in the form of a gel that loses its consistency in the interstitial space and reaches the free liquid state inside vessels. Water therefore is the ideal agent for communication among the different organs, cells and proteins—most of them are enzymes.

All enzymatic reactions take place in water, which transfers the conformational change information from enzyme to reactants and product nearby. In the inactive stage, this communication is shielded by the dielectric effect of water on local ionic fields.[7] Following an enzymatic signal, which could be a calcium ionic pulse, the frequency of the collective ionic field becomes high enough to overcome the water's shielding effect. With the electrostatic shield removed, the strong oscillations of the enzymatic protein, resonating with the collective ionic field, are transmitted to the solvated reactants and solvated product, where they coordinate the chemical interactions between them. It is these polarization oscillations that, under water control, drive the fundamental process by which chemical reactions occur.[7,16]

Furthermore, water has the capacity to dissolve a remarkable array of polar molecules that serve as fuels, building blocks, catalysts, and information carriers. By transporting these substances all over the body through its network, water is, in the saying of Whyte, the "supreme non-speck facilitator of chemical change."[10]

Water as agent of retardation of physical velocities

Life, in essence, is a manifestation of organized energy, a time-requiring process. For water to be at its foundation, it must possess a special property that allows for such energy organization to occur. Although a generally accepted theory of space and time is not available, we know that events in the high-temperature physical world occur at astonishingly high velocities. Taking Plank's time as basic reference, physical events can occur every 10^{-43} second, a timeframe infinitesimally too small to be conceived by human minds.

Several steps in living structure contribute to the slowdown of physical velocities, but the first and most crucial one is that imposed by the dipole moment of water molecules. The biochemical reactions in the water phase occur at speeds ranging from femtoseconds to microseconds, many orders of magnitude slower than those of physical processes.[17]

During evolution, the delay brought about by the thermal inhibition of motion allowed for information to be processed and structure to be self-organized. A new impulse for energy organization was thus generated, not at high temperatures and fast velocities but at low temperatures and slow velocities. It was based on molecules, not atoms. And the basic molecule upon which all the biological edifice was ultimately built was that of water.

SELECTED REFERENCES

1. KANDEL R. Water from heavens. Columbia Univ. Press, New York (2003).
2. MENTRE P, HOA GBH. Effects of high hydrostatic pressure on living cells: a consequence of the properties of macromolecules and macromolecule-associate water. In. Rev. Cytol. 201, 1-84, 2001.
3. SAENGER W. Structure and dynamics of water surrounding biomolecules. Ann. Rev. Biophys. Biophys. Chem. 16, 93-114, 1987.
4. WATTERSON JG. Model for a cooperative structure wave. In: Biophysics of water. F. Franks and S. F. Mathias (eds.), Wiley, N.Y., 144-147 (1981).
5. WATTERSON JG. Does solvent structure underlie osmotic mechanisms? Phys. Chem. Liq., 16, 3-316, 1987.
6. WATTERSON JG. What drives osmosis? J. Biol. Phys. 21, 1-9, 1995.
7. CARVALHO JS. Life: its physics and dynamics. American Literary Press, Inc., Baltimore, MD (2003).
8. LUDWIG R. Water: From clusters to the bulk. Angew. Chem. Int. Ed. 40, 1808-1827, 2001.
9. REID BL. Further appreciation of a control system for chemical reactions residing in virtual energy flows through the bio-system. Med. Hypoth. 52, 227-234, 1999.
10. WHYTE LL. The unitary principle in physics and biology. The Cresset Press, London (1949).
11. WATTERSON JG. The interactions of water and proteins in cellular function. Prog. Mol. Subcell. Biol. 12, 113-134, 1991.
12. WIGGINS PM. Role of water in some biological processes. Microbiol. Rev. 54, 432-449, 1990.
13. DROST-HANSEN W, SINGLETON JL. Liquid asset—How the exotic properties of cell water enhance life. The Sciences 29, 38-42, 1989.
14. URRY DW. Elastic biomolecular machines. Scient. Amer. 272, 64-69, 1995.
15. POLLAK GH. Cells, gels and the engines of life. Ebner & sons, Seattle, WA, USA (2001).
16. MCCORCKLE RA. A physical basis for biochemistry. J. theor. Biol. 148, 393-400, 1991.
17. AXELSSON S. The basic reality of mind and spongiform diseases. Med. Hypoth. 57, 549-554, 2001.

CHAPTER 4

THE UNIVERSAL PROCESS AND TIME FLOW

The unity of man and nature, of subject and object,
appears to be grounded in the unity of time.
C.F. von Weizsäcker, 1971

Space and time are the two components of the matrix of the universe. They are so intimately related that special relativity does not distinguish them—the spacetime concept. Time, however, has a unique property: it flows. But the flow of time is not like that of a river. It is made up of a chain of separate events like the ticks of a clock. Between ticks there is no time.

Time is a profound but treacherous concept: its nature is not established and, under certain circumstances, its reality may not be manifest. There are various facets to time, but here we are concerned with the direction of time flow in natural systems and processes. In them, it has been assumed, time can either flow forward and backward reversibly or it can flow irreversibly in the forward direction—the arrow of time.

Historically, the first measurement of time was based on the motions of our planet. The earth's rotation on its axis provided the units of the clock (day), and the earth's orbit around the sun provided the units of the calendar (year). Later, in the seventeenth century, Newton derived his laws of motion from the planetary trajectories of the solar system. These laws do not distinguish forward time from backward time, so they imply time to be reversible.

But Newtonian dynamics does not agree with our experience of reality. For us, there is no doubt that time is irreversible. Spring does not turn back into winter and people do not get younger. In metabolism, the assimilation of oxygen and food always proceeds in a manner which can never be reversed. There must be some assumption in Newton's mathematics that is not valid for our observable universe and for organisms.

Time became more complex with relativity but Einstein continued to consider it to be reversible. The special theory further implies that temporal relations are fundamentally and generally similar to spatial relations. To complicate matters even more, quantum mechanics and modern string theories are also rooted on the reversibility of time. At a fundamental level, it does appear that physics is time-symmetric. Saying it another way, the laws of physics are indifferent to the direction of time.

Over the years, however, not all physicists have shared the time-symmetrical view of the universe, but the apparent contradiction has not been resolved. The so-called 'time paradox' continues to be an obstacle to a most desired unification of physics and biology and to an eventual development of a general theory of science. Most importantly, the time stalemate has blurred the connection of physical nature to human nature.

In retrospect, Newton studied a very simple and stable system—the solar system. Its components are sustained by gravitational forces and move across a medium of very low density. With constant, low-level entropy (a measure of change) and virtually no friction, the planets follow predictable trajectories. Since the orbits look as sensible in one direction as they do in the other, time is seemingly reversible.

Applied to systems of similar behavior, Newton's gravitational laws are as true today as ever. But they cannot be used to describe the dynamics of many-component complex systems, such as the heat-producing organisms operating far from thermodynamic equilibrium. According to the nonequilibrium scientist Ilya Prigogine, massively interactive systems are highly sensitive to initial conditions. A slight deviation in the path of one component induces effects so unpredictable that the new and old trajectories of the system have nothing in common. In systems behaving this way, time is obviously irreversible. For Prigogine, time irreversibility emerges from the dynamic instability inherent to complexity.[1,2]

An even more fundamental explanation, not from dynamics but from atomic processes of energy exchange, was that advanced by the theoretical physicist Lancelot Whyte in the second quarter of last century. For Whyte, it is the electrical and radiation processes that are essentially irreversible. Radioactivity, for instance, is only observed in the form of disintegration and not also as the reverse process of spontaneous building up of heavier elements from lighter ones. It is the irreversible motion of light and the time delay in light perception that account for our experience of time irreversibility.[3]

Logically, if an arrow of time is admitted to exist, it could be conceived that *all physical laws are irreversible*. In this scenario, the apparent reversibility of the gravitational laws of motion would be seen as a limiting case, a result of the negligibly small irreversibility of the solar system.

Whyte went on to develop a general theory capable of describing the essential features of cosmic and protoplasmic organization.[4] Since irreversibility implies motion, the static concept of matter of classical physics, which implies permanence, could not be used. And, as all physical experience requires a certain amount of time, relativistic time based on instantaneous coincidences was also inadequate. New concepts of matter and time had to be formulated.

A new conception of matter and time

The concept of static matter was replaced by that of dynamic *structure*. In some new way, structure links matter with energy. Structure is continually changing, and given the order of nature this change is not arbitrary. It must obey to some rule of logic. It was the cause of this ordered change, the theoretical equivalent to some hidden 'force' moving structure, that Whyte set out to discover. To achieve this dauntless task, his theory should possess the highest possible degree of comprehensiveness and simplicity.

Obviously, once the concept of structure was accepted, the classical concepts of matter and energy as separate entities were discarded. But the conservation of matter and energy forms the basis of classical thermodynamics, one of the pillars of modern science. The new theory, therefore, should be more fundamental and capable of explaining all other existent theories, including thermodynamics.

Furthermore, to properly represent the fundamental structure of physical phenomena, the theory should be largely descriptive. It should be capable of describing not only the topological facts (environment, surface, interior, shape, etc.) but also the past history of the universe and of organisms, including sudden occurrences such as impact of meteorites, volcanic eruptions and cell division.

This required a local form of metric based on sequence of events rather than motions of celestial bodies. Whyte chose a time measurement derived from space measurement, a temporal relation known in physics as relation of *succession*. And in order to represent irreversible processes, the *asymmetrical* relation of succession (an event B is *later than* an event A) was required.[5] This asymmetrical relation vanishes in the limiting (symmetrical) case when event B and event A can be regarded as simultaneous.

In this way, time became a dependent variable. And, as in the case of matter and energy, time as an independent entity or an extended quantity was discarded. This seemingly unimportant local or biological time, not to be incorporated into the laws of motion, would profoundly change the understanding of our relations with nature.

Introduction to the unitary principle

Structure at large can be conceived as a system of ordered relations among all components. These relations can be of diverse nature, but for science it is the *causal* or space-time relations—of size, shape, position and orientation—that are most important. Therefore, throughout the theoretical treatment, the principles of *causality* and *symmetry* were applied.

To approach natural conditions as much as possible, the relations of *inequality of cause and effect* (A ≠ B), *temporal succession* and *spatial asymmetry* were chosen. They are more comprehensive than their symmetrical congeners of classical physics—equality of cause and effect (A = B), reversibility, and spatial symmetry—and can include them as limiting cases.

On theoretically analyzing the characteristics of the causal relations in isolable one-way processes, in which cause and effect were not equal, Whyte was struck by the possibility that later states could be more symmetrical than earlier ones. Some other workers, notably Pierre Curie, had already arrived at similar conclusions, but only a prepared mind looking ahead for specific novelty could recognize the potential importance of such a fact.

According to the principle of causality, isolable processes must display internal causal continuity (absence of arbitrary features) one-way in the temporal succession, from earlier to later states, and not necessarily the other way round. So, on logical grounds, an increase in symmetry in later states must correspond to a *decrease in internal asymmetry* in some earlier states. This reasoning was the turning point in the scientific development of what turned out to be the most general theory of process ever conceived.

In applying theoretical concepts to reality, the causal continuity of scientific theory, which relates earlier to later states in a process, was taken as *tendency* towards a definite end-condition. Tendency to symmetry became the axiom of a causal science of one-way process. This axiom, which Whyte called *unitary principle,* can be expressed in just a dozen words:

"Asymmetry tends to disappear, and this tendency is realized in isolable processes."

By asymmetry it is meant an observable deviation or distortion of a system from a specific type of symmetry, either rotational or translational.[4]

Tendency to symmetry is the *cause* of structural change. Continuity in the sequence of change is assured by the order linking earlier to later states. This ordered continuity is called *process.* Since the above process displays only one form of continuity, from asymmetry to symmetry, it is called *unitary.*

An *isolable process* or system (these terms are interchangeable) is one evolving toward isolation and static symmetry. This evolution is continuous but may comprise several isolable steps of process, each consisting in the separating out of a characteristic symmetry (Fig. 4, pg. 54). If an existing symmetry disappears in the course of any process, that process is not isolable, and must be treated as a component of some more extensive process. This is also true if a characteristic asymmetry does not wholly disappear and the correspondent latent symmetry never becomes explicit.

Contrary to some systems (crystals, some molecules) where complete isolation is reached (or nearly so) and no further change is possible, living systems are too complex and so interlinked with their environment that no complete separation of symmetrical form is possible. They remain, through mutual adjustments, in a state of compromised equilibrium with their environment.

Unitary theory remains descriptive, i.e., nonquantitative. Quantitative treatment would require the expression of our physical constants and measurements in *relational terms,* which are more general than quantitative terms and physically more significant. Instead of the classical dimensions of mass, length and time, the quantitative theory would require the use of relative lengths, angles or angle variables which are more appropriate to the description of stable structure.

This would call for a geometrical atomic theory, which is unavailable at present, and for precise knowledge of the primary particles of the field. For these reasons, the mathematical formulation of a fundamental theory of microstructure, based directly on the changing spatial relations of physical entities cannot be undertaken at present state of physical knowledge. This fundamental theory will be the future *theory of exact science* predicted by Whyte.

The unitary theory is a scientific formulation derived from an intrinsic property of structure that treats living systems as ordering processes. It will need considerable more mathematical work and requires development when further knowledge of biological and mental structure is at hand. But the theoretical concepts are sound and contain in themselves the potentiality for further growth.

The theory has been disregarded for a long time, but when its implications are properly understood, it will change drastically our current views on biological organization and will give us a new understanding of the world. Its biological concepts, however, must be validated before they can be accepted. In this regard, the structural patterns predicted by the theory should be amenable to confirmation by already available magnification techniques. The form of the theory could also be modified if a concept of change more fundamental than asymmetry to symmetry is eventually found.

The *unitary field and its process*

Space is not empty. It is filled with some unknown stuff we call the field. The field is not amenable to direct observation, so its conception must be theoretical, either with or without mathematical foundation. Each space or space-time theory, therefore, has its own conception of field.

The unitary field is unique among all the described fields: space is changed from a field basis to a structured basis. The field is a structured entity which represents the process of a system rather than the state of a medium linking separate structures. Particle and field are indistinguishable. In simple terms, it is a field of polarized structures, and the phenomenon is nothing but the changing structures and their relationships.[4] The field provides within itself the potentialities for structure, process, organization, and one-way development.

The ultimate structures of the unitary field are not known but they manifest a spontaneous tendency towards structural symmetry (through changes of size, shape and position), which is the source of the *unitary field process*. This tendency involves two aspects, functionally inseparable from one another: (i) tendency towards a decrease in asymmetry (or an increase in symmetry), and (ii) tendency toward uniformity of structural asymmetry in the field as a whole. This uniformizing tendency is known as the *normalizing process*.

The universe can be considered an unstable system of systems where no system is completely isolable, or isolable for too long, in all respects. By itself, each system tends to perfect its inner symmetry and, as part of more extensive systems, it tends to conform to the symmetry of its neighbors. As a whole, the field tends to maintain uniformity of asymmetry. It is this normalizing of the field that maintains the universe in process. Otherwise, over time the finite structural asymmetry of the universe would be vanished.

In the realm of living organisms, the two aspects of the field are in close cooperation via the synchronized anabolistic and catabolistic processes, the balance being in favor of the latter. The catabolistic process is the normalizing process in a highly differentiated form. It is highly sensitive to environmental variables that ultimately direct the development of biological organization at all levels of the hierarchy.

In a supportive environment, the field tendency from asymmetry to symmetry can be associated with an all-important phenomenon: development and transformation of structure. The atoms, stars and galaxies were formed that way. So were molecules, macromolecules, life and thought—all the structure existent in our galaxy and in the whole universe. On this view, stars, humans and all things built by humans are no more than huge aggregates of unitary field structures of differing complexity. The unitary field process is the common link which relates everything with everything else.

Physical structure was build up from a vast number of copies of a few fundamental components? electrons, quarks, gluons, and photons. These are also the elementary building blocks that physical and biological systems are made of. Organic structure was formed from physical structure under the influence of the natural forces and the guidance of the unitary process. The appearance of life on earth was simply a straightforward consequence of a series of structural transformations. Molecular and biological evolutions are thus connected to the general evolutionary process of the universe.

The measurement of time

The physical conception of time arose from the practical utility of clocks for describing natural processes. The history of time measurement attests to the difficulties in devising a reliable clock with an accuracy that could satisfy both astronomers and physicists. This was only achieved with the discovery of the atomic clock.

To measure time, some quantitative process is required. Two such processes have been adopted, and these define two independent, fundamental measures of time: (i) *epoch,* which specifies the moment when an instantaneous event occurs in the sense of time of day, and (ii) *time interval* or duration of a continued event. Under the present system of time measurement, epoch is stated in terms of *mean solar time* and time interval in terms of *atomic time.*

It was astronomers that first defined time in terms of the rotation of the earth about its axis. More refined variants of astronomical time then followed, but another rotational time, the mean solar time, came to be finally adopted. From January 1, 1972 onward, the Greenwich Mean Time, the mean solar time of the longitude of Greenwich, England (taken as zero), became the basis for *standard time* throughout the world. Our *local time* is the time for our particular meridian of longitude, one of twenty-four standard meridians which give rise to twenty-four standard time zones.

But the speed of the earth's rotation is not constant. It is slightly irregular, and therefore it does not provide the best standard of time. A further complication arises from the need to have a whole number of days in the year—365, with 366 in leap years. Thus, once in a great while an additional second must be inserted, a sixty-first second in a particular minute, to keep pace with atomic time. In spite of its drawbacks, rotational time remains in common use because it is of necessity for civil purposes, navigation, geodetic surveying, and space-vehicle tracking.

In the 1960s, the physicists introduced a measure of time interval in terms of atomic time, which later became the *fundamental unit of time.* Today's atomic clock is based on the cesium-133 atom when irradiated with the microwave frequency of 9,192,631,770 hertz ($\sim 10^{10}$ oscillations/sec). This chosen frequency corresponds to the energy difference between two specific internal energy levels of the cesium-133 nucleus. The output signal of the atomic clock is fed back to the frequency oscillator to prevent it from drifting from the resonance frequency. In this way the device is locked to this frequency with an accuracy which at present is of 1 part in 10^{15}, or 1 second in thirty million years.[6]

Dividing the output electronically by the number of oscillations per second provides one-second 'ticks' of the atomic clock: the *SI unit of time.* Greenwich's time scale is adjusted by introducing one second, a leap second, whenever the astronomical and physical times diverge by more than 0.9 second.[7]

Physical time vs. psychological time

The high-energy primary particles of the physical world have not been identified and so their oscillation rate is not precisely known. If, however, the electromagnetic Planck time is a reliable measure, the interval between oscillatory events at fundamental level is of the order of 10^{-43} human-constructed second. This is equivalent of saying that there are 10^{43} *physical events*, or clicks of Planck time, in one human-constructed second.

The enormity of this exponential becomes more apparent when this gargantuan number is related to another whose dimension is easier for us to conceive. Such a number can be the estimated age, in seconds, of our galaxy: less than 10^{18} human-constructed seconds,[8] which corresponds to less than 10^{28} atomic clock oscillations.

We can now see that the whole galactic time, measured in terms of *physical events*, can be squeezed in less than one human-constructed second! This simply means, as Axelsson rightly pointed out, that at field level time does not exist.[9] There is no *physical time*. By human standards, all fundamental events are instantaneous.

In unitary theory, time does not exist when the field is in its asymmetrical state. It begins to exist when the field displays its intrinsic asymmetry to symmetry tendency. When the state of maximal (static) symmetry is reached in a particular system, time stops. On this view, time is a manifestation of the unitary principle.[10]

Fundamentally, time is a property of the unitary structured field. In a given isolable system, physical or biological, a more symmetrical state is necessarily later in time than a less symmetrical one. Tendency from asymmetry to symmetry points the direction of events in time—time's arrow.

In unitary terms, each click of physical time may be conceived as the motion of two oscillating primary particles as they approach one another to momentarily form a more symmetrical quantum particle. Given the extremely high physical velocities, this event occurs instantaneously.

The field physical velocities were considerably slowed down in atomic structure, by lengthening the interval between clicks of physical time, particularly at the level of the outer-shell electrons.[11] Further slowdown, however, was required for complex organic structure to be assembled. The first molecular structure facing this intimidating task was that of water, the matrix of the organic realm. The time delay provided by the water network was enough to allow enzymatic catalysis to occur, which in turn further slowed down physical velocities. Then, with increasing structural complexity, from bacteria to man, the span between physical clicks became more and more extended.

In the human organism, the temporal aspects of structural organization are openly manifested and have a special role in the central nervous system, the most complex of all organic structure. They are at the basis of all our perceptions, memories, and anticipations and of all temporal functions of the mind. These mental patterns are thus different from the structural patterns of all other tissues where spatial relations are more important.

Central to the temporal nature of mental processes is the continuous self-assembly of quantum structural aggregates from individual, more-asymmetrical quantum field particles, in some unidentified stratum of the brainstem reticular space. According to Baranski, each aggregate takes about 0.1 sec to form.[10] There are thus ten such events in one human-constructed second. This is the pace of *psychological time.*

The formation of one specific quantum structural aggregate corresponds to one click of biological time. Time perception, however, depends on the continuous formation or run-off of sequences or chains of structural aggregates as the normalizing process moves through the reticular space. It is the spatial ordering process of the aggregates into sequences that is most directly connected to our perception of time[10].

Each quantum structural aggregate is not a statistical ensemble but a more structurally symmetrical transactional form of numerous, more-asymmetrical quantum field particles. The unitary flow of sequences of structural aggregates thus corresponds to a quantum field superposition of coherent states. Unitary theory is a modified, one-way quantum field theory.

Psychological or biological time should not be confused with the hypothalamic biological "clock," which uses changes in light intensity transmitted from special retinal photoreceptors to control the activity of behavior and metabolism. This so-called clock does not set biological time. It sets circadian biological rhythms. All rhythmicity, circadian or otherwise, is related to local fields generated by planetary motions and is not fixed in the structural pattern of the unitary field.

SELECTED REFERENCES

1. ROTHMAN T. Irreversible differences. The Sciences 37, 26-31, 1997.
2. PRIGOGINE I, ANTONIOU I. Laws of nature and time symmetry breaking. Annals N.Y. Acad. Sci. 879, 8-28, 1999.
3. WHYTE LL. Archimedes, or the future of physics. E. P. Dutton & Co., New York (1928).
4. WHYTE LL. The unitary principle in physics and biology. The Cresset Press, London (1949).
5. WHYTE LL. One-way processes in physics and biophysics. Brit. J. Philo. Sci. 6, 107-121, 1955.
6. WYNANDS R. The atomic wrist-watch. Nature 429, 509-510, 2004.
7. KLEPPNER D. An instant in time. Nature 410, 1027, 2001.
8. PASQUINI L, BONIFACIO P, RANDICH S, GALLI D, GRATTON RG. Beryllium in turnoff stars of NGC 6397: Early Galaxy spallation, cosmochronology and cluster formation. Astron. Astrophys. 426, 651-657, 2004.
9. AXELSSON S. Perspectives on handedness, life and physics. Med. Hypoth. 61, 267-274, 2003.
10. BARANSKI L. Scientific basis for world civilization. The Christopher Publishing House, Boston, USA (1960).
11. WILCZEK F. Whence the Force of F = ma? II. Rationalizations. Physics Today 57, 10-11, 2004.

CHAPTER 5

THE CELL AS UNIT OF LIFE

The cell is the only unit of biological organization
that can multiply itself.
Lancelot L. Whyte, 1948

Life is not autonomous. Living systems are connected to the physical world, and it was from this connection that life was born. Because of its dependency on molecular reactions, life required an environment of its own, a compartment where structure could be self-organized and functionally integrated, under the guidance of the unitary field. For that to take place a boundary was required.

The cell membrane and the membrane of cell organelles may have developed in porous, clay-like mineral grains of the earth's regolith, as soon as liquid-drop water became available. The process may have started before the oceans were created and then continued in the bottom of the oceans.

Clay particles are thin mineral sheets with high surface charge density. Their thickness is measured in nanometers but their length and breadth reach the micrometer-millimeter size range. They pack by aligning in parallel array to form stacks which can become millimeters high. These clay stacks closely resemble the stacked lipid bilayers of chloroplasts, mitochondria and the Golgi.[1]

Presumably, the primitive cells were assembled in microcavities within these mineral grains, which functioned as natural non-cellular biochemical reactors. Water molecules first diffused into the virtual space between two flat clay sheets, forcing them apart, and from there into the many pores traversing the sheets. Over a long stretch of time, other molecules then followed—purines, pyrimidines, amino acids, simple fatty acids, micelles, carbon compounds, phosphates, etc.—and became adsorbed onto the surfaces.[2] Some of these compounds were delivered onto the juvenile earth's surface from space, but the majority was synthesized locally.

RNA-like molecules, synthesized by the catalytic activity of the mineral surfaces, could have served as informational matrices for the synthesis of oligopeptides. Adsorbed micelles, converted into vesicles, may have then incorporated RNA, oligopeptides and other substances, including loose pieces of catalytic mineral. A virus-like biological structure with a lipid envelope was eventually formed. When it reached a critical size, it was expelled from the grain to continue evolving independently.[2,3]

In the absence of direct information on the beginning of life, the scenario pictured above, although plausible and supported by experiment, cannot be proven. Nevertheless, this elementary open system, isolated from the environment and far from equilibrium, could have served as the basis for the creation of the very first biological system.

The operation of the unitary field

Hidden within the grain of clay, there was an essential element operating continuously towards the assembly of the above hypothetical system. This hidden force, not to be found in fossil records, is the same force that have guided the formation of atoms, stars and galaxies. It is an expression of the universal structured field. Only this field contains, in potentiality, the impulse and order required for the building up and organizing of structure. The unitary field was the central force, the *primum movens* of evolution.

The final conception of the unitary field has not been written, since the conditions for the presence of the different physical fields, in terms of geometrical relations in three-dimensional space, have not been defined. Nevertheless, the unitary field contains within itself the electromagnetic field of classical physics.

The decrease of electrical polarization by the mutual attraction and neutralization of electrical charges of opposite sign is interpreted in unitary theory as a decrease of a *polar asymmetry*. But unitary theory carries the polarization concept further: polarization tends to a minimum, and polarization differences in ultimate structures also tend toward a minimum.[5]

The same can be said for magnetic influences, which in unitary concepts become *axial asymmetries* or rotational distortions. It is, therefore, still correct to say that life has evolved from the electromagnetic field. Only the concepts have changed, not the reality.

To be very responsive to electromagnetic fields, living structures ought to be easily polarizable. They must behave like dielectrics.[4] Gels, polymers, and colloids are examples of such structures, and they are also the major components of biological systems. Living structure is, of necessity, of soft and very delicate nature.

Under the influence of the unitary field, biological structure, and most particularly *protein structure*, manifests very characteristic behavior. It has the propensity to self-organize and to automatically adapt to environmental variations. These properties are conferred to structure as component of the one-way developmental process of the field. If structure were to be taken out of this specific field, it would lose all those properties.

The first biological organism was able to evolve because it was propelled to do so by the mysterious creative field that pervades the universe. Without the impulse of the field no life would ever be possible. But in the dawn of evolution, the environmental variations were constantly distorting the structuring and organizing of the field. Further evolution could not pursue without adjustment of the developing structure to its environment, a process we call *adaptation*.

The unitary properties of living protein

To get an insight into the adaptive process which took place in the early stages of life, we must describe first the special characteristics and dynamical properties of proteins in the developed animal cell. As we saw in chapter 3, protein dynamics is intimately connected to that of adjacent water and ions. Living protein, therefore, refers to the protein-ion-water system.

Proteins are assemblies of amino acids held together by strong covalent bonds, which form the backbone of the protein and gives stability to it. This peptide backbone has a large permanent dipole in the peptide plane made up of numerous CO-NH dipoles.[4] The side and end groups of the constituent amino acids are maintained by much-weaker, non-covalent connections, which gives proteins considerably more freedom of movement, allowing them to fold and unfold as required. Most of these lateral groups have either charges or permanent dipoles in them, and they are the main source of instability.

All charges and electrical dipoles, asymmetrically located in three-dimensional space, interact with each other to form an intrinsic electric field around the protein. Each protein domain may have its specific field. It is in the protein field that structural asymmetry, the thermodynamic equivalent of free energy, is stored, and it is by way of this tiny field that the universal unitary field exerts its influence on living proteins.

We can think of these electrical dipoles as miniscule batteries, capable of charging and discharging millions of times per second in a pulsating fashion. Dipoles are not only present in proteins and nucleic acids, but also in other biopolymers such as polysaccharides and in water. They constitute the basic electrical components of functional structure. Since life is electromagnetic in nature, any disturbance of whatever kind affecting living systems is ultimately transmitted to, and compensated by, changes originating in these electric dipoles.

The dipoles, therefore, have a cyclic or pulsating function, a back and forth type of movement. In their normal state, they are maintained in a *metastable* condition of very low threshold. Charge is maintained by a distortion of the structure of the protein, a deviation from the equilibrium state equivalent to the stretching of a spring. In this position, the dipoles are polarized.

Following a given stimulus, which in unitary terms corresponds to the induction of a polar asymmetry, the system threshold is overcome and spontaneous depolarization (relaxation) occurs. The depolarized system or part of a system is transformed into a more symmetrical form. The released structural asymmetry is used in whatever function is required and triggers induction of structural asymmetry from an extended neighboring region and/or from local metabolic sources and from the environment. According to unitary concepts, the environmental component carries field asymmetry.

The resultant extended field is a normalizing field that repolarizes the exhausted dipoles, brings the polarizing gradients to a minimum, and adjusts the level of polarization of the region to conform to that of the field (normalizing process). In so doing, it spreads order by aligning all the dipoles in a specific way.[5]

The property of chirality

The physical, chemical and biological properties of a molecule are dependent not only on the nature and number of the constituent atoms but also on the position of these atoms in three-dimensional space. A minority of molecules spread in space in such a way that if bisected by an axis or plane, the two halves are identical and mirror images of each other. Objects, such as spheres and cubes, butterflies, and even symbols, such as the letters E and H, also behave this way. They are said to be bilaterally symmetrical.

Nevertheless, most molecules in nature do not manifest this property. They are asymmetrical, and a special case of asymmetry, not to be confused with the asymmetry that the unitary theory refers to, is chirality (chiro, from the Greek *cheir,* the hand). The chirality element can be an axis or a plane but is more commonly a center; that is, an asymmetric atom.

A molecule is chiral if does not possess an internal plane of symmetry and the molecule and its mirror image are not superposable (by rotations). Examples of objects belonging to this type of chirality are hands, screws, keys and spiral staircases. Both our hands have a thumb plus four more fingers in the same position but they are not identical. The right hand does not fit properly into a left-hand glove. Chemists refer to mirror-image molecules as L-enantiometers and D-enantiometers, L and D standing for levo (left) and dextro (right).[6]

In many organic molecules, however, chirality results from a characteristic of the carbon atom, which arises from the ability of this atom to form tetrahedral structures. A central carbon bonded to four different groups or atoms, situated at the corners of a tetrahedron, introduces asymmetry or chirality because, under these circumstances, two spatial forms, an R-handed and an L-handed, are possible, one being the mirror image of the other. The molecules of many substances in nature, such as the common sugars and other carbohydrates, contain more than one asymmetrical carbon atom.

Of all the naturally occurring amino acids in proteins, only glycine (CH_2-NH_2-COOH) has a plane of symmetry along its 'spine' and thus it is achiral. All the other twenty one genetically encoded amino acids so far known are chiral, having L- and D-enantiometers. For some strange reason, only *L-amino acids* were selected during evolution to be constituents of proteins.

Furthermore, in their most common form of secondary structure, the polypeptide chain of proteins is coiled in such a way as to form a *right-handed* alpha helix. This configuration is required for proper catalytic activity, presumably because of its stability. Synthetic chains of L- and D-amino acids do not twist in the way necessary for efficient catalytic activity—they cannot form the alpha helix.[6]

The nucleic acids RNA and DNA are also chiral-selective. Each is composed of four types of subunits, each of which incorporates a chiral sugar group, but in this case the D-enantiometer was the one selected. Like proteins, RNA and DNA also form *right-handed* helices, but in this case with the exclusive presence of *D-sugars*.

So, the chemistry of life is chiral and based almost exclusively on L-amino acids and D-sugars. The ability of biological molecules to discriminate between enantiometers is therefore vital for living systems.

Chiral asymmetry extends throughout all scales, from the fundamental particles to atoms, molecules, and galaxies. A general theory of chirality is still unavailable, but it appears that the physics of terrestrial biopolymer chirality may be different from,[7] and even unrelated to,[8] that of the physical systems. Perhaps the hypothesis recently proposed by He et coworkers may shed some light on the biological chiral phenomenon.[9]

Chirality is attributed to a field generated by the planetary motions, mainly those of earth—its rotation around the spin axis and its revolution around the sun. Both are right-handed helical motions that produce a natural, rhythmic and chiral, right-handed helical force field over the earth in space and time.

According to the above hypothesis, the axis of helical R-handed biomolecular enantiomers tends to be parallel to the axis of the helical force field, so less intrinsic energy is required for stabilizing R-handed helical structures. This is the reason why terrestrial living systems selected both R-handed nuclei acids based on D-sugars and R-handed proteins based on L-amino acids.

The theory also explains the biorhythms, circadian and others, of biological systems on the earth. These rhythms are fundamental control systems that regulate a wide variety of biophenomena, such as behavior, sleep and activity pattern, development, metabolism, gene expression, aging, etc.

Independent of the origin of chirality, the chiral information is now stored in the hereditary units of plants and animals. Regarding life, the chiral process starts in basic synthesis carried out by green plants and special bacteria, where the central symmetry of isolated carbon and nitrogen atoms is converted into the linear or cyclic patterns of carbohydrate and protein molecules.

We must assume that the atomic pattern of the protein molecules guiding the process of basic synthesis possesses a unique primary axis, vector or not, and some kind of spiral process about that axis for it to produce the chiral patterns and specific handednesses found in the synthesized organic compounds.[5]

Chiral processes, therefore, must involve long-range multi-term relations, which the short-range atomic and molecular interactions of classical biochemistry are unable to achieve. Only a field of polarizable structures and polarization axes guiding and controlling the shape, orientation and position of the parts involved, as proposed by unitary theory, can achieve so specific relationships.

Structure, process and complexity

In classical dynamics, when a system cannot be treated as being isolable, it is considered to be in interaction with another system. The interaction affects only the relative motion of the two systems, which are taken as unchanging entities. This kind of picture is, at best, a limiting case of the unitary view.

In unitary theory, if a system is not isolable, it is considered to be a part of a wider system in process. The parts are continually changing entities which are related to the whole in the following ways:

(i)   Each part has an intrinsic tendency to develop its characteristic symmetry. However, this tendency is modified by the overriding process of the whole. The processes of the parts must adjust to (cooperate with) that of the whole for a single result to be achieved: the development of the symmetry of the whole.

(ii)   The parts (units of process) may interact, not only in pairs but in groups, which may be arranged in a hierarchy of sub-groups within the complete system (whole), each group or sub-group having its own tendency to symmetry.

(iii)   The formative process may involve not only changes in the spatial relations of the parts (external processes) but in their atomic pattern as well (internal processes).

(iv)   The relations of whole to parts are determined by the states of polarization of the whole and of the parts.

Unitary theory gives us a different view of reality. Structure is continually changing, and therefore its components cannot be identified. In effect, structure is not easily differentiated from process. We can only know structure when static symmetry is reached and structural transformations end. Structure is a limiting case of process.

In classical dynamics, interactions are based on the concept of *force*. They always occur in equal and opposite action-reaction pairs between unchanging bodies. Clearly, some other concept rather than force must be used to describe changing structure. Whyte has proposed the concept of *order,* since the multiple relations making up structure must be ordered relations.

For *order* to be achieved, and maintained against continuous thermal and minor mechanical disturbances, a certain size of structure is required. Biological systems characteristically contain a very high number of particles or degrees of freedom of kinetic theory—they are *complex systems*. To describe complex systems in constant change a *theory of process,* such as that proposed by Whyte, is required.

Complex systems are built from simpler ones by the addition of parts in a continuous process of structural transformations. But the adding on of parts is a self-limiting process. When a certain degree of complexity is achieved, the behavior of a system loses regularity and becomes unpredictable and chaotic and no further build up of structure is possible. A new structure must be started de novo. So, biological organization is *hierarchical* in nature.

Organic stability

Viewing the ground from a landing airplane, we first see a somewhat indistinct pattern, which becomes more and more revealing as altitude diminishes. In this situation, altitude is equivalent to the inverse of magnification. As the distance to the ground decreases, more magnification is achieved.

Like physical structure, biological structure has a characteristic pattern under each level of magnification. If were possible to magnify a structure high enough, we should be able to see its ultimate atomic *pattern*. This new concept of pattern underlies that of *form* or, to say it another way, form is grounded in the ultimate pattern. For biology, form, as the overall expression of the particular arrangement of atoms in a given complex system, is more important than its material components, which are devoid of individuality.[10]

Under the unitary view, process is best understood as the *development of form*, in appropriate selected systems. To initiate form development, a *threshold* must be passed and some structural asymmetry must be released from a given structure. The latter becomes isolable from the process of the environment and spontaneously relaxes into a more symmetrical form. This transformation involves a change in the internal atomic pattern of the structure. After a threshold, a new pattern separates itself out. Whyte calls the formation of new structural patterns by a decrease of asymmetry the *formative process*.

In the process of new structure formation, a point is reached where the residual asymmetry of an isolable system is so low that the creation of new patterns of lesser asymmetry is no longer possible. The system has reached a *stable state*. Full symmetry, however, has not been achieved. The system is still in a state of minor internal distortion, and an asymmetry gradient still exists between the system and the environment. Under these circumstances, and by the continued influence of the unitary tendency, the pattern of a stable structure will tend either to multiply or extend itself, if this is possible. Thus, the unitary tendency to full symmetry works not only internally but externally as well —*translational symmetry*.

For the purpose of this discussion, we may assume that the system above described, which could be a synthetic enzyme, has reached the state D of the theoretical development of full symmetry schematically represented in Fig. 4 (pg. 54). If the environment is appropriate, the system can still reproduce its pattern by facilitating the synthesis of a new identical pattern (multiplication) or by imposing its pattern on that of a less stable neighboring system (extension). The complete process of the unitary tendency is schematically represented in Fig. 5 (pg. 55).

The fact that more stable patterns are *dominant* over those less stable is a consequence of the unitary process and a direct result of the relations of parts to wholes. When a stable pattern (+++) and a less stable one (XXXXX) come into proximity and a whole is formed, the asymmetry of the whole will tend to decrease, by lessening the distortions of the two components, according to the unitary principle. For equal decreases of distortion, the higher threshold of the more stable pattern protects it from change (+++), and it is the less stable and more asymmetrical one that ends up being transformed into a new pattern that conforms more closely to that of the more stable component (+X+X+). The more stable pattern dominated the less stable by spreading its pattern into it. More stable patterns enjoy ascendancy over less stable ones, and their own stability tends to be reinforced.

Protein multiplication is a complex autocatalytic process carried out in little steps, the details of which are still obscure. One of these steps is the induction by the field of the specific pattern of the prototype protein molecule. Imprinted in the pattern of each protein, and therefore in its resultant field, is a *record* of the process by which it was formed, and it is this record or memory which leads to the local synthesis of a new identical molecule. Stable protein with specific patterns, such as genes, enzymes and other catalysts, can be looked upon as printing presses. And as such, they can generate as many copies as necessary.

The process of catalytic multiplication of specific structures or chemical units and the still more complex process of organic synthesis do not occur in isolation. The evocation of synthesis or multiplication, the local synthetic or multiplicative process and the resultant products and their effects are all components of a single comprehensive normalizing process.

Unitary theory gives us a different view of a living cell. Processes occur automatically in a simple and ordered manner. Polarized structures and respective fields work in perfect harmony. Proteins need not be transported through the cytoplasmic gel, from ribosomes to cell membrane or to other target sites, as present organic chemistry asserts. They can be synthesized in situ, by copying atom by atom, a duplicative process endowed with extreme reliability.

The extraordinary stability of species character throughout the generations, and of tissue character in the growth of the individual, attests to that reliability and strongly suggests that the multiplication of stable unit patterns, genes and enzymes, underlies normal organic synthesis.

In adult life, the physical stability of organic structure is due to the structural stability of the molecular patterns of its constituent proteins. Structural patterns that distort the normalizing process are selectively excluded simply by denying structural asymmetry to them. But it is the *asymmetrical relation* between more stable and less stable parts, or between a stable system and a less stable part, that is the fundamental factor underlying all development and maintenance of pattern, and consequently all biological organization.[5]

It is well recognized that organic stability is a requirement for life. Steve Grand, one of the proponents of artificial life, considers the most important law of nature to be: "What persists, persists, and what don't, it don't."[11] This "law," however, is just a consequence of the unitary principle.

Unitary theory in its present form is just an outline of a mathematically based future formulation capable of explaining all other theories, and thus of unifying all sciences. Although its schematic form is not appropriate to explain detail, it is powerful enough to explain other theories in general terms.

Watterson's theory attributes organic stability and order to the packing of spatially compatible units: water clusters, protein domains and other structural quanta of similar size. The theory is based on classical reversible physics and uses the concept of matter and the *gas law* to energetically link solids (proteins) to gases via the water cluster. Clusters are maintained by inwardly directed tensional forces derived from hydrogen bonding. They are packets of *structural energy*. Larger clusters, such as those of solvent water, have more energy, or structural asymmetry of unitary theory, than the smaller clusters of solutes.

Watterson has created a new "field" of mesoscopic particles: water clusters and, by similarity of size, protein domains. This molecular or biological field is made up of stable particles. Stability is achieved by the action of tensional forces that, at pixel-size level, are at a minimum. Tension can spread laterally by the orderly addition of particles that become united by a single tensional force. For a particle to increase in size, energy is required to overcome the basic tension. The enlarged particle can then spontaneously decrease to pixel size with release of structural energy.

For unitary theory, tension is just a particular form of structural asymmetry. In both theories, new form and energy result from structural transformations. Watterson's theory, however, is not based on process, and the cluster concept is not directly derived from the basic physical field.

As theory of process, unitary theory is based on one-way physics and on the concept of structure in constant change. Structural asymmetry is not contained in isolated clusters, but it is an integral part of structure that, in isolable systems, tends to spontaneously slide into symmetry. To go back from symmetry to asymmetry, an input of structural asymmetry is required. In living structure, structural asymmetry is maintained at a certain level by the normalizing process of the field.

In terms of size, there are no quanta in unitary theory. The special concept of polarization allows parts to be of various sizes. Each part has its own formative process, which is dominated by the formative process of the whole and so all parts cooperate toward the symmetry or stability of the whole.

So, in Watterson's theory we have a hierarchy of sizes; in unitary theory we have a hierarchy of formative processes sustained by relations of dominance. In one theory, structural stability is based on tension and structure is organized from the bottom up; in the other, structural stability is based on symmetry and structure is organized from the top down. The unitary theory is more general and can include Watterson's theory because tensional deformation of structure is just a specific way of generating structural asymmetry, albeit one of very high biological importance.

The changes in water structure brought about by the interplay of protein surfaces and adjacent ions, so beautifully described by Wiggins, can also be explained by unitary concepts. These changes occur at the molecular level and continue to take place within water clusters. In unitary theory, any structural transformation must involve a change from asymmetry to symmetry. High-density, vapor-like, water is more asymmetrical than the low-density, ice-like, water. Under isolable conditions, high- density water spontaneously converts into low-density water. Then, structural asymmetry must be inducted, for example by ATP hydrolysis, into low-density water to bring it back to the high-density state. Asymmetry-symmetry changes are continuously occurring. They are at the basis of the living processes.

The unitary view of evolution

The fundamental process in evolutionary development was *facilitation* of the intrinsic creative tendency of the normalizing field by the developing organism. In the dawn of life, with no proteins to supply the required organic stability, evolution was threatened by the constant disruption of the normalizing process by environmental variations. Some mechanism had to be developed to constantly adjust the evolving organism to its changing environment.

According to Leo Baranski, a unitary theorist contemporary of Whyte and an authority in the energy field, a molecule that had an essential role in this adjustment process was ATP, a nucleotide synthesized from readily available building blocks. The ATP molecule consists of the purine adenine linked to D-ribose and three phosphate groups, bound together linearly by covalent bonds. Two of the phosphate groups possess what are called *energy-rich phosphate bonds* due to the large quantity of "free energy," or structural asymmetry, they carry.

ATP was able to transport structural asymmetry and, by way of its terminal phosphate group, to induct it into other compounds without heat loss. This ability was recognized by the unitary field which used ATP as its agent on the molecular level. In other words, the tiny ATP system was the primitive normalizing process, inducting structural asymmetry and directing it to the development of structure.

Besides the activity of asymmetry induction, the nucleotide ATP also possessed the activity of catalysis. Both of these properties enabled the ATP molecule to become the precursor of a very important macromolecular entity, a deoxy-D-ribonucleotide polymer, which grew in time to become DNA.[12]

Today's human DNA is an immensely long threadlike structure in the shape of a twisted ladder. The two sugar-phosphate units, the molecule's backbone, form the vertical pieces, and the nitrogenous base units form the rungs of the ladder. The sugar-phosphate chains are coiled around a common vertical axis, giving the structure a right-handed, double-helix configuration. One full turn of the helix is completed within the vertical length of 3.4 nanometers, which corresponds to ten rungs of the ladder.

The helical shape of the sugar-phosphate chains imposes a special order on the sequence of the nitrogenous bases forming the ladder rungs. Each rung is made up of two out of four possible bases, the base pair always being a purine (adenine (A) or guanine(G)) and a complementary pyrimidine (thymine (T) or cytosine (C)), connected by a hydrogen bond. Steric restrictions and hydrogen binding requirements allow only pairs of A and T or G and C to fit into the two-nanometer-long space between the vertical chains.

The bases are distributed in irregular order. Reading vertically along a side piece of the ladder we may find a base sequence similar to this: GTTCATTTACGT... and so on and so on. In this apparently meaningless string of bases, a code of biological order was found: a sequence of three consecutive bases, called a *codon*, is translated into one amino acid (p.e., the codon GTT contains the DNA code for valine, in most organisms). With a four-letter alphabet, a total of sixty-four three-letter words are possible, enough to encode the primitive twenty amino acids.

The length of the entire human DNA ladder is huge on the atomic scale. It comprises about three billion rungs! If stretched out, it would extend for almost a meter. There are two such DNA ladders per cell, one from each parent. To fit all this DNA material into a nucleus of no more than ten μm in diameter, a high order of packaging is required.[13]

To the effect, each long double stranded DNA fiber is divided into twenty-three unequal lengths, corresponding to as many different chromosomes. There are thus twenty-three pairs of chromosomes per cell. To organize the packing, the delicate linear DNA segment of each chromosome is wrapped around cores of positively charged proteins, called histones, to form a tightly bound nucleoprotein structure—the *chromatin* fiber. After several levels of coiling and super coiling, the linear chromatin fiber is finally shaped into a compact three-dimensional wad. This packing arrangement is highly regulated[14] and anticipates the replication of chromosomes at the time of cell division.

The coded regions comprise only 2 percent of the total DNA molecule! Within chromosomes, they are scattered in groups or units of base sequences called *genes*. Usually, each gene or *hereditary unit* encodes the chemical structure of a single protein. In the human genome, about 25,000 genes have so far been found. Only the bases of the DNA molecule carry genetic information, their sugar and phosphate groups performing only a structural role.

At present state of knowledge, we can only give a very broad and fragmented account of the basic unitary processes which presumably took place in early evolution. Since our concern is the hypothetical animal cell from which the human species has evolved, all other ramifications of evolutionary process will not be mentioned.

The assembly of the nucleoprotein structure of chromosomes was initiated early in evolution. Nucleotides were already available as well as amino acids, which were linked in a long chain to make up histones. The ATP-nucleoprotein system then supplied the evolving organism with the potentiality of being differentiated by the environment.[12]

As agent of the normalizing process, ATP continually carried to the base structure of the nucleoprotein patterns with particular configurations related to external environmental conditions. These patterns, made of quantum field structures, represented records or memories of the environmental variations. Later, patterns of similar configurations were aggregated, hierarchically organized and internally transformed into the symmetrical and polarizable structures known as *genes*. They were to become the agents of *biological order*.

Continually cycling between ATP and nucleoprotein nuclei, the polarization pulses of the primitive unitary process continued their organizing activity. After a long stretch of time, special circumstances led to the assembly, in the highly polarizable aqueous and electrolytic matrix of the cytoplasm, of a long chain of amino acids in a unique and highly specific order. This chain was then folded in a special conformation to become the first *protein enzyme*.

The polarization level of this new organic structure conformed to that of the field defined by ATP and the nucleoprotein. For the first time in the history of our planet, structural asymmetry was directly transmitted, by way of field quanta, to the dipoles of a protein structure to start a new organization, which was to lead to what we now call life.[12]

Using this enzyme as prototype, other identical or similar enzymes were synthesized by the inductive action of the normalizing field. These enzymes, now acting as agents of the normalizing process, led to the synthesis of patterns of structures via long sequences of physico-chemical reactions. These patterns were then organized by relations of dominance into more complex structures, which facilitated normalization on a large scale. Although patterns were different, the normalizing field assured that all structures shared a uniform level of polarization.

Evolution, however, was constantly facing new challenges that had to be solved if further organizational development was to continue. With the synthesis of more and more protein and other structure, the cytoplasmic gel starts forming and primitive motion was ready to appear. The gelled systems became more sensitive to changes in the environment, which meanwhile also increased its sources of normalizing distortion. The field alone could no longer supply the required structural asymmetry, and the asymmetry level of the organism started decreasing.

To prevent the symmetry level from falling to a dangerously low level, the evolving organism had no other choice than import from the environment, and later from the sun, the needed structural asymmetry. To this effect, the normalizing process slowly developed enzymes which formed cyclic processes (oxidation-reduction) for the oxidation of reduced minerals and extracted from them structural asymmetry. These were the first *catabolistic systems*.[12]

Utilizing the biological order in the DNA system, the normalizing process continually produced more differentiated and increasingly organized structure, which in turn became increasingly sensitive to environmental variations. More specific enzymes were then synthesized, and more efficient cyclic processes were formed. Prominent among those was the oxidation-reduction of gluthatione, described by Holt.[15] Powered by anaerobic glycolysis, this cycle is capable of amplifying energy by generating one extra electron per cycle.

However, the energy crisis was not resolved until energy extraction from sunlight by green algae became possible, followed by the gradual development in the animal cell of a more sophisticated catabolistic enzymatic process, the *mitochondrial apparatus*, for oxidation of the photosynthetic products. The organism was now able to increase the elaboration of its structural organization and to reproduce itself without the danger of decreasing its asymmetry level.

Meanwhile, the external membrane of the evolving organism became more organized. In addition to the original lipid assembly, it acquired an array of stable extended protein structures and enzymatic respiratory systems of its own. Together with DNA, the cell membrane took over the overall normalizing process of the cell.

Cell as biological unit

The two major parts of a cell are the nucleus and the cytoplasm. The nucleus is separated from the cytoplasm by the nuclear membrane, and the cytoplasm is separated from the surroundings by the cell membrane. The cell also contains many organized physical structures called *organelles,* each enclosed by its respective membrane. All membranes are composed primarily of lipids and proteins.

According to electrical theory, polarization occurs in biological systems whenever there is separation of ionic charge between the sides of a membrane. A gradient of polarization or potential difference is thus established which can be sustained if the membrane possesses certain intrinsic differential permeabilities to the ions involved.

In the case of the cell, for instance, K^+ and impermeant organic anions are present at higher concentrations inside the cell than outside while the opposite is true for Na^+ and Cl^- ions. This uneven distribution of ions inside and outside the cell, together with the contribution of active Na^+/K^+ pumps, generates an average potential of 60-90 mV across the cell membrane, cell side negative.

In unitary theory, the concept of polarization is deeper than, but not inconsistent with, that of electrical theory. In unitary terms, a polarized system is not one where there is a separation of electrical charge. It is one that displays an asymmetry which can be represented by a direction with sense. Since the conditions under which polarization is measurable have not yet been determined, the magnitude of polarization is provisionally ignored. Charges are regarded as components of a polarized structure.

As a quantum field theory, unitary theory involves propagation of quantum field particles. These particles are polarizable three-dimensional structures. Therefore, the theory does not distinguish between waves and particles. In unitary concepts, waves and particles are the same thing.

Instead of dealing with charges directly, unitary theory deals with polar asymmetries, such as electrical dipoles, where all charges are included. The polarization field of the cell is produced by the electrical dipoles of all biopolymers, mostly proteins.

But unitary theory goes further. In any system, polar asymmetries tend to decrease towards a state in which the polarization disappears. If, for some reason, this is not possible, there is a tendency to reduce polarizations differences to the minimum permitted by the internal and external conditions of the system. This tendency is called the *normalizing process.*

To define a normalizing process, two boundary conditions are required. In the case of the cell, the external boundary corresponds to the cell membrane. It is a variable boundary through which the cell and the environment mutually interact. It stabilizes and varies the internal boundary which corresponds to the hereditary units within the nucleus. The hereditary units are stable structures, akin to *organic crystals.* They are unchangeable, although mutations sometimes occur.

Under normal conditions of operation, a gradient exists between the cell membrane and the hereditary units, between the outer boundary and the inner boundary. This gradient is in constant change. The organism is continually stimulated by, and adjusting to, its environment. It is this gradient that is normalized. If the gradient disappears, life cannot be sustained and death inevitably follows.

One-way pulses of polarization are continuously going from the membrane to the hereditary units and from the hereditary units to the cell membrane (Fig. 6, pg. 56). In addition to this *extended normalizing process,* countless local normalizing processes are constantly in operation.

Only one extended normalizing process controls the whole cell at a given time. The normalizing pulse which organizes a specific cell can be generated in the membrane of that cell or propagated to it from a group of neighboring cells from which the specific cell is a component.

For this reason, a cell cannot perform an isolated internal function and a differentiated function at the same time. For instance, a cell cannot divide itself and transport products or propagate polarization at the same time. At any moment, it can either divide or perform a differentiated function.

It is the normalizing process characteristic of each cell that allows the cell to organize itself after division of the parent cell. Only a cell, and not other structure bigger or smaller than a cell, can organize itself. This makes the cell the *biological unit,* the basic block for organic construction. Seen this way, the cell may be considered the *organic galaxy.*

In multicellular organisms, each differentiated cell has its own normalizing process, which is coordinated with the normalizing processes of other cells by differentiating links in the areas of functional continuity. In each cell, the nucleus is the region of undifferentiated multiplication characteristic of the species; the membrane is the differentiated boundary linking the cell with other differentiated cells and with the environment; and the cytoplasm is the field of interplay of heredity and environmental factors.[5]

Whether the organism is unicellular or multicellular, the cell membrane constitutes the boundary separating it from the local environment. This separation, however, is far from complete, since the cell ultimately gets its nutrients from the environment and discards heat and catabolic products into it. The organism appears to extend to the environment or, saying it in other terms, organism and environment form a single system.

The single system *organism-environment,* however, does not share the same constancy of conditions. The cell membrane allows some control to be exerted on the cell interior against disruptive environmental variations—thermal, mechanical, electromagnetic. In this way, cell conditions remain reasonably stable, the fragile living gel is not excessively stimulated, and structural order is maintained within a narrow range.

SELECTED REFERENCES

1. WATTERSON JG. The pressure pixel—unit of life? BioSystems 41, 141-152, 1997.
2. HANCZYC MM, FUJIKAWA SM, SZOSTAK JW. Experimental models of primitive cellular compartments: encapsulation, growth, and division. Science 302, 618-621, 2003.
3. NUSSINOV MD, MARON VI. Impulse paradigm of self-organization of matter in the universe, and nanobiological principles. Nanobiology 2, 215-228, 1993.
4. CARVALHO JS. Life: its physics and dynamics. American Literary Press, Baltimore, MD (2003).
5. WHYTE LL. The unitary principle in physics and biology. The Cresset Press, London (1949).
6. HEGSTROM RA, KONDEPUDI DK. The handedness of the universe. Scient. Amer. 262, 108-115, 1990.
7. CAPOZZIELLO S, LATTANZI A. Spiral galaxies as enantiometers: chirality, an underlying feature in chemistry and astrophysics. Chirality 18, 17-23, 2006.
8. BONNER WA. Parity violation and the evolution of biomolecular homochirality. Chirality 12, 114-126, 2000.
9. HE YJ, QI F, QI SC. Effect of chiral helical force field on molecular helical enantiometers and possible origin of biomolecular homochirality. Med. Hypoth. 51, 125-128, 1998.
10. WHYTE LL. Accent on form. Greenwood Press, Publishers. Westport, Connecticut (1954).
11. GRAND S. Creation: Life and how to make it. Weidenfeld & Nicolson (2000).
12. BARANSKI LJ. Scientific basis for world civilization. Unitary field theory. The Christopher Publishing House, Boston, USA (1960).
13. VON BAEYER HC. Information. The new language of science. Harvard Univ. Press (2004).
14. MOHD-SARIP A, VERRIJZER CP. A higher order of silence. Science 306, 1484-1485, 2004.
15. HOLT JAG. Some characteristics of the gluthatione cycle revealed by ionizing and non-ionizing electromagnetic radiation. Med. Hypoth. 45, 345-368, 1995.

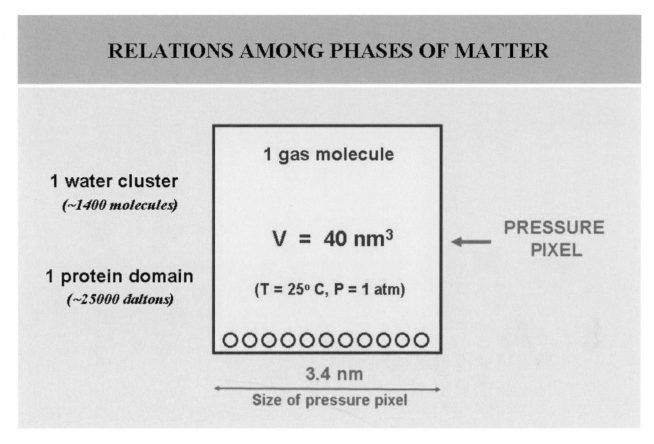

Fig. 1. **Schematic representation of the concept of water cluster.** The volume occupied by 1 gas molecule is equivalent to that of 1 water cluster and of 1 protein domain, at equilibrium and under conditions of normal pressure and temperature. Theoretically, this volume corresponds to a cube of 40 nm³ volume and 3.4 nm sides. It is the minimal volume that sustains pressure in a liquid (pressure pixel). Changing the shape of the volume does not invalidate the concept. Further explanation in text.

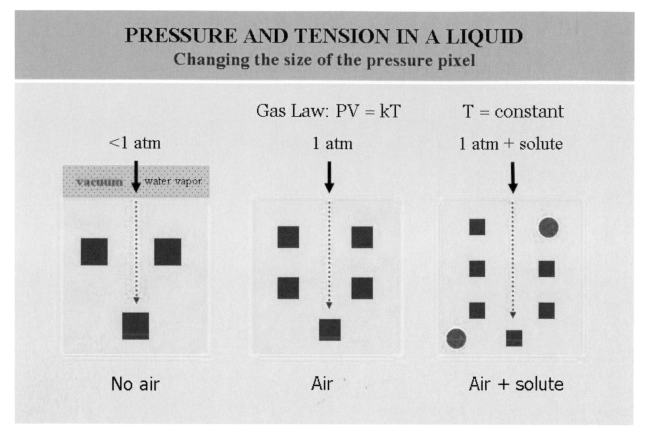

Fig. 2. **Schematic visualization of the effect of pressure changes on size and concentration of water clusters, at constant temperature.** The figure shows three beakers with water. The first on left is exposed to negative pressure, the second to normal pressure and the third to normal pressure plus dissolved solute. The water clusters are represented by the blue cubes and the solute by the red circles. With increasing pressure, cluster size diminishes and cluster number increases. Pressure is exerted on the outside of the clusters and is equilibrated by inside tension. Addition of solute has the same effect as increase in pressure. Air molecules dissolve in water but they do not exert additional pressure as solutes do. Further explanation in text.

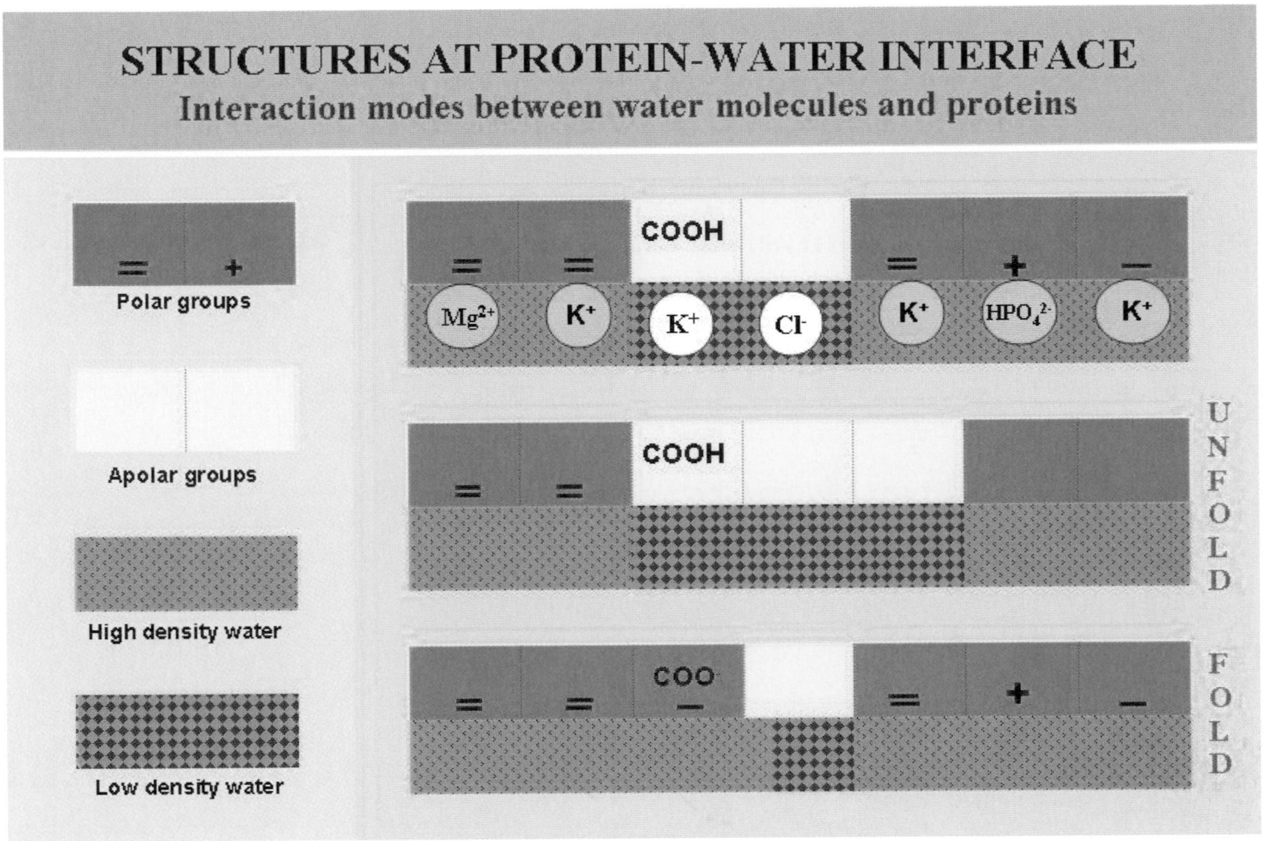

Fig. 3. **Very simplistic representation of a polymer transition, from the unfolded to the folded state.** The *upper panel* represents a segment of a polymer side-chain made up of polar (charged) and apolar (uncharged) regions. Water molecules arrange themselves around these regions in different configurations, and counter-ions tend to accumulate in water as indicated. Potassium ions, the most abundant of the intracellular cations, bind preferentially to the fixed negative charges of the polymer, because they are excluded from high-density water. Ion distribution is asymmetric but the water network must ultimately equilibrate at the same chemical potential.

The chemical conditions under which folding takes place depend on the mix of polar and apolar regions in the side-chains. If the polymer's side-chains are mostly apolar (hydrophobic), a great amount of low-density (stretched) water must be destructured before the chain can contract. The polymer stays in its extended configuration *(middle panel)*. Removing hydrophobic groups or converting them into hydrophilic ones, which exert a strong pull on water molecules and disrupt nearby low-density water, forces the polymer to contract *(lower panel)*. It has been shown that side-chains of glutamic acid can take the form of COOH or COO-, depending on local pH. Further explanations in text.

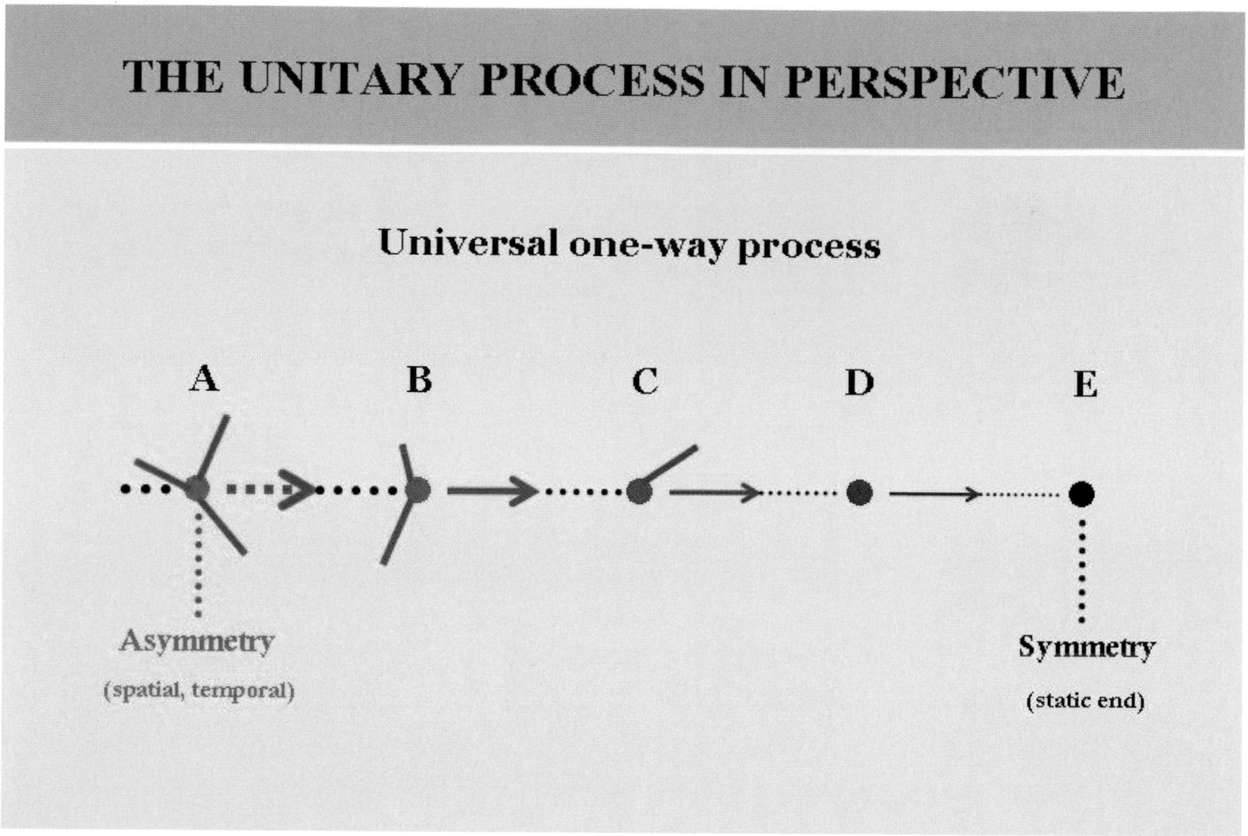

THE UNITARY PROCESS IN PERSPECTIVE

Universal one-way process

A   B   C   D   E

Asymmetry
(spatial, temporal)

Symmetry
(static end)

Fig. 4. **Mental representation of a material system on the way to complete isolation (without the interference of external factors).** In the system represented (●), there are five states (A,B,C,D,E) and four one-way units of process (A-B, B-C, C-D, D-E), each leading from a threshold (causality) to a terminus (finality), where the asymmetry characterizing the unit of process vanishes. The distortion due to the wider system of which the system represented was previously a part is progressively eliminated as the system in question separates itself out and perfects its characteristic latent symmetry. Symmetry is the limit that is approached as the degree of isolation of the system increases. The stepwise loss of structural asymmetry (release of energy) between the initial and final state is represented by the direction and thickness of the arrows. To reverse the direction of the sequence, induction of structural asymmetry into the system would be required. Further explanation in text.

THE COMPLETE UNITARY TENDENCY OF A SYSTEM

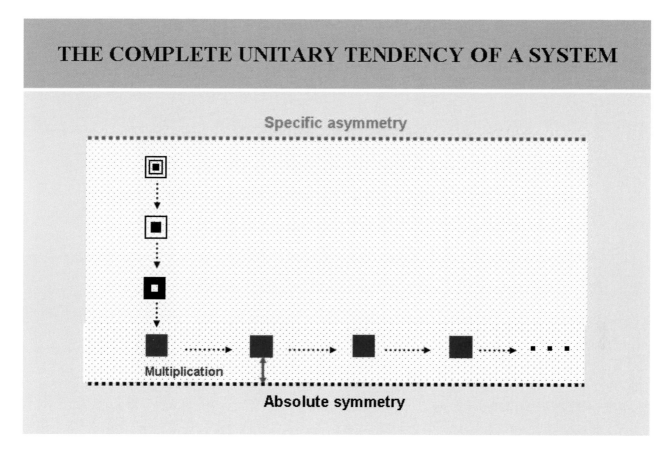

Fig. 5. **Schematic representation of the total process of decreasing asymmetry in a biological system.** For simplification, the wider system of which this biological system is a part is ignored. Structural patterns of decreasing internal asymmetry develop in an isolable system, under the influence of the unitary tendency. This development process culminates in a stable pattern of characteristic symmetry. If the environment is appropriate, the residual asymmetry of this stable pattern tends to extend the pattern. The stable pattern is simply inducted, in part or whole, on neighboring structures by the normalizing field, and this autocatalytic process of multiplication is easily repeated if necessary. The unitary tendency of a system, therefore, operates not only internally but externally as well. Further explanation in text.

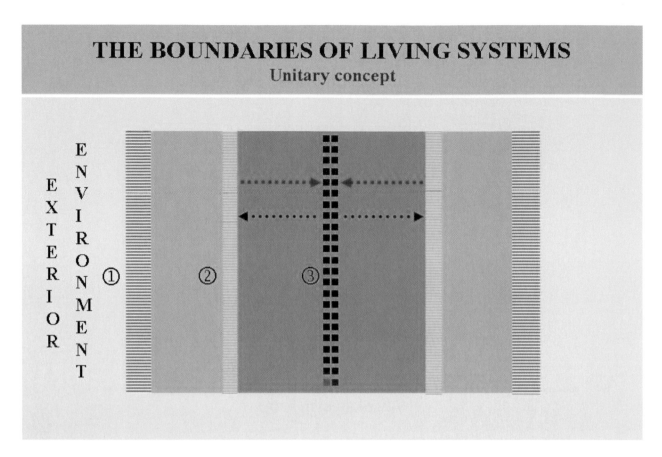

Fig. 6. **Perspective representation of the functional boundaries of a living system.** The boundaries are indicated by numbers. 1 – boundary between exterior environment and living system (variable, large changes); 2 – boundary between interior environment and cell (variable, small changes) or outer boundary of cell (cell membrane); 3 – inner boundary of hereditary units (stable, unchangeable). The black arrows are meant to illustrate the direction of the field stabilized by boundaries 3 and 2, which carries information from the hereditary units to all parts of the cell and to the environment; the red arrows illustrate the extended aspect of the normalizing process. Boundary 2 is the external limit of the continuous normalizing process, through which the environment and the system mutually influence one another.

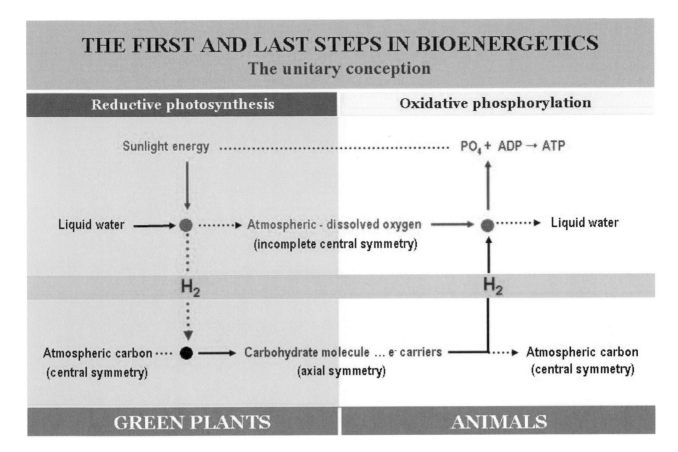

Fig. 7. **Schematic view of the fundamental steps involved in the membrane processes of reductive photosynthesis and oxidative phosphorylation.** In photosynthesis, photonic energy splits water molecules and the resultant hydrogen atoms are used to convert the point symmetry of separated carbon atoms into the axial symmetry of carbohydrate molecules. In oxidative phosphorylation, activated oxygen inducts structural asymmetry (transfers free energy) into the symmetrical but less stable fuel molecules, which are fractured in the process. The structural asymmetry is first used to build up a proton motive-force (proton gradient). Dissipation of this force then transfers the asymmetry to ATP, where it is converted into a phosphoryl potential. The initial reactants in photosynthesis appear as end-products in oxidative phosphorylation. These processes are complementary and form between them an overarching water-carbon-oxygen cycle. A more detailed account of the pertinent biochemical pathways is given in ref. 12.

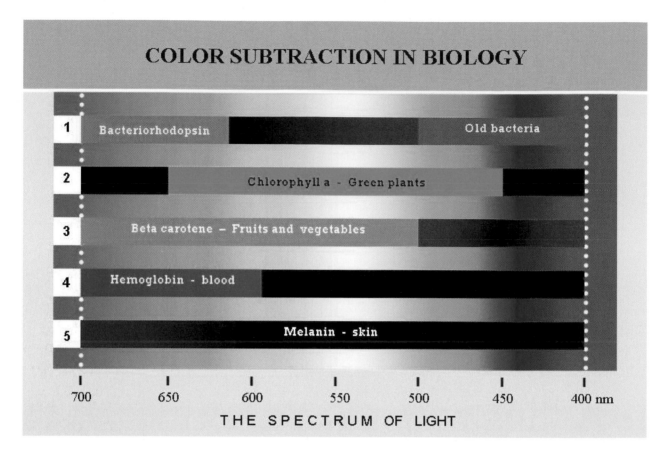

Fig 8. Schematic representation of light absorption spectra and colors of five organic pigments. The light electromagnetic spectrum is pictured on the background. Absorptions bands and colors are approximate. **1.** Bacteriorhodopsin is a primitive photosynthetic pigment developed by *Halobacterium halobium.* It absorbs light in the middle of the spectrum and for this reason has a purple color. Its chromophore (light-absorbing group) is retinal, an aldehyde derivative of vitamin A and a constituent of our visual pigment rhodopsin. **2.** Chlorophyll a is one of the photoreceptor pigments responsible for the green color reflected by most plants. It contains a network of alternating single and double bonds that strongly absorb light-energy in two bands located at each extreme (red and blue) of the visible spectrum. It is this absorbed energy that drives photosynthesis. The basic structure of chlorophyll molecules is a magnesium-containing porphyrin, related to hemoglobin. **3.** Beta-carotene, a compound containing eleven conjugated C=C bonds, is the source of the orange color of carrots. Its orange hue is a sign that it absorbs light in the blue-violet end of the spectrum. **4.** Human hemoglobin comprises two pairs of polypeptide chains, with each chain folded to provide a binding site for a hem group, a porphyrin with four nitrogen atoms holding an iron (II) atom as a chelate. Human blood looks red because hemoglobin, contained in the red blood cells, reflects light in the red part and absorbs it in the yellow and violet parts of the spectrum. **5.** Melanin is nature's sunscreen. It is a dark brown or black pigment that reabsorbs ultraviolet light (not shown) as well as visible light all over the spectrum. The molecule, presumably a polymeric protein, is produced by melanocytes in skin, iris, retina, hair, and elsewhere. In the skin, the melanocytes are packaged in structures called melanosomes, which are taken up by keratinocytes to shield their DNA.

CHAPTER 6

THE ANIMAL ORGANISM

*Organisms are dynamic material systems that constantly
reproduce their own material organization.*
Peter Cariani, 2001

With the creation of the cell, the first level of organization, the *biochemical level,* was finally achieved. The unicellular organism acquired the properties of stability, adaptation, self-regulation and division, which enabled its survival and multiplication. Under challenging environmental conditions, it also had the potentiality for furthering its own development.

It was an astonishing, almost incomprehensible achievement for the normalizing field to organize structure from isolated simple molecules to complex integrated wholes displaying so amazing properties. Judging by the time dispended in the process—over three billion years—it must have been an unusually challenging undertaking.

But the cell was to be the foundation of the edifice of life, a sophisticated new realm that should be capable of processing information, making decisions and enjoying some autonomy in relation to the environment. The cell, as basic unit, should be built to support those activities.

Once water was selected as matrix of life, the processes of the field had to exploit all the properties this medium could offer to realize those functions. Water requirements must have imposed constraints on the patterns of structural proteins that were possible, on the settings of thermoregulation, and on the propagation of polarization throughout the whole hierarchy. To achieve the high degree of coordination exhibited by the cell, the one-way developmental process ought to be necessarily slow.

To surprise us even more, the structural elegance of the cell is contained in a space just of a few cubic micra. The dimensions of an average cell are fifty times below the minimum our eyes can distinguish. We could easily fit one hundred of them on a head of a pin. But in its tiny cocoon a cell contains hundreds of thousands of proteins, most of them enzymes, and many other structures. It is a very crowded place but not a disorganized one. Within a cell a special type of order prevails.

Some unicellular organisms, such as bacteria, adjusted to their environment so successfully that their form has remained with little change over the years. Given their fast multiplication, measured in about a half-hour, bacteria have grown to astronomical figures. Today, they are everywhere: in water, on land, in air, outside and inside us. Truly, this is the "age of the bacteria", as Gould puts it.[1]

In spite of being small in size and limited in the quantity, quality, and differentiation of their protein structure, bacteria also possess a normalizing field, all internal components being in perpetual process. They are capable of basic synthesis, some specific synthesis, growth and multiplication, and some can perform movements in a relatively uncontrolled manner. They are not necessarily less complex than other multicellular organisms but are certainly more heterogenous. We may not find two proteins with identical patterns and similar neighbors in them.[2]

Under the continuing influence of external factors, some primitive unicellular organisms, in the sea and on land, continued their differentiation, and after millions of years of constant adaptation and DNA development they became *multicellular*. During this long period, the organisms drained the environment of energetic substrates and replaced them with waste products of their catabolism. With ever-meager resources available, closed recycling loops were developed in which one life source's waste became another's food.[3]

With multicellularity came the differentiation of organ systems and glands and the appearance of special communication systems of vascular and nervous origin. And, after five hundred million years of continuous development, a second organizational level was reached, the *biological level,* which may be divided into physiological and perceptual levels.

The arrival of the mammals was followed by further development of the central nervous system, and another five hundred millions years later, after several warm and cold cycles of adaptation and major geological catastrophes, the social *Homo sapiens* gradually emerged. In the course of the fifty thousand years or so since his appearance, a third and special level of energy organization, the *cognitive level,* has continuously been evolving.

Over the entire time of evolution, the relation between organisms and environment was intimately intertwined. Organisms modified environment, and the modified environment, in turn, modified the organisms. Active life and passive environment continuously co-evolved. A blatant example of this seesaw evolution, on the large scale, was the rise in atmospheric oxygen by the primitive photosynthetic green algae, which led to the subsequent development of aerobic respiratory pathways by animal cells.

The interrelation of organism and environment continues to occur in modern times, not always for the better. We are adapting to an accelerating rise in the carbon dioxide level in the atmosphere, brought about not by exaggeration of natural causes but by the added oxidation of fossil fuels for the convenience of man. The pressure of dissolved carbon dioxide in our blood has been increasing since the start of industrialization. If the atmospheric CO_2 rise continues at the present rate of 1.16 p.p.m. annually, our capacity for full compensation may be exhausted by 2050.[4] Before that happens, and if new mechanisms of adaptation do not evolve, the depth of our respiration will be forced to increase, which, among other things, will impose an extra burden to our hearts.

We are a product of the basic field, but what maintains us alive are the stimulations emanating from our environment, to which we must constantly adapt. The system organism-environment is a part of the wider system of our environment. Both are maintained in a dynamic and delicate balance. Any change affecting the processes of the environment will inevitably affect the processes of the organism.

Organisms vs. networks and mechanisms

A superficial look at the therapeutic methods utilized by modern medicine, namely the transplantation of organs from one human being to another and the use of artificial "organs" to replace the function of diseased or missing ones, conveys the impression that man is a machine made of independent, and therefore static, components. This view of living systems is supported by classical physics, which is based on concepts of permanence of matter and conservation of mass/energy.

But man is not a machine made of invariant parts obeying already established physical laws. The animal organism is an organized system of systems in perpetual process. Unlike a machine, which is made by man, or by other machines made by man, an organism is self-constructed. It obeys natural laws, which may not be the ones of current physics theory.

The man-as-machine concept is a product of quantitative physics. In his quest for the understanding of nature, man used numbers in the process of measurement. Current physics and the scientific method of investigation are based on them. But structure, inorganic or organic, does not contain abstract numbers. It is a system of relations, asymmetrical relations, obeying a certain type of order. The overall order of things is what physical observation perceives.

By stressing the asymmetrical relations of structure and admitting an orderly one-way process underlying evolutionary development, Whyte has given us a clear differentiation between organisms and mechanisms. The science of quantity is unable to provide concepts and symbols such as those of structure, order, development, form, thought and quality, which are characteristic of thinking organisms. In the description of structural patterns, it is order, not quantity, that is most important. Quantitative relations can only express restricted aspects of the observable order.

There is a tendency nowadays to compare organisms, or systems within organisms, to physical networks of links, nodes and hubs. Some of those, such as telephone exchanges and the e-mail internet, are simple connection networks where the average number of connections per node changes little throughout. This type of network can grow unconstrained (it is called 'scale-free') because there is no need to rapidly integrate information or globally respond to the current state of its nodes. Living structure is not organized this way.

More sophisticated networks do exist, such as stock exchanges and computer systems, where operation relies on the interrelated activity of any or all its component nodes. In this type of network, the number of connections per node must increase faster than linear as the network grows. Under these circumstances, if nothing is done, connection saturation (or prohibitive cost) eventually occurs, followed by decline in network efficiency.[5]

In principle and in practice, the increase in global connectivity required by functionally organized systems ultimately imposes limits to their growth. When connection limits cannot be raised or functional components cannot directly communicate with each other, *hierarchies* are introduced for the purpose of increasing efficiency. These hierarchies are called management in business organizations, control systems in engineering, and regulation in current biology.

In organisms, complexity also results from an increase in the number of components, and their spatial order is also hierarchical, but the nature of the components and the relations between them are fundamentally different from those of man-made networks. There are no inert nodes in living structure. Instead, there are field-generating polarized parts with an intrinsic tendency, themselves components of a comprehensive one-way normalizing process.

The connections or interactions between nodes correspond to the asymmetrical relations of parts to parts and parts to wholes, described in the previous chapter. If process is what sustains an organism, there is no room in it for "interactions" between two unchangeable entities. In their tendency to symmetry or in their adjustment to the symmetry of their wholes, the parts are always internally transformed in some way. Structural transformations, induced by polarized fields on polarizable parts, are at the basis of all processes in organisms. Furthermore, the transfer of structural asymmetry from a field to a part is a *vectorial process*, which inducts a certain position and orientation into the structure of the part. Structural order is ultimately born from information transmitted by specific polarized fields.

Under the unitary principle, the isolable phase of the process of a complex system consists in the decrease of some asymmetry characteristic of the system with the appearance of a new more symmetrical form. The ultimate structural units of process are thus formative groups. These groups may again be grouped to form a higher system characterized by its dominant process. In organic systems, the ultimate units are arranged in a hierarchy of sub-systems within the complete system. Organisms are, at bottom, not hierarchies of scales but hierarchies of formative processes.

The precise structure and general properties of living hierarchies are still unknown. They are viewed as three-dimensional units where some or all of their first-order parts are subject to delicate and intricate constraints on their degrees of freedom. The properties of these units are not simple linear summations of the individual properties that these units would display if isolated.[6] Hierarchies with these characteristics are only possible in systems of low energy and attenuated physical velocities, such as organisms.

A recognized property of organic hierarchies is their capacity for generating heterogeneity, which is so characteristic of living structure. Classical close-packing can only be associated with homogeneity. This heterogeneity, maintained by the asymmetrical relations between levels, may be at the basis of the intrinsic stability manifested by this type of architecture.

Contrary to non-biological hierarchies, which are imposed on systems from without, natural hierarchies are generated from within the systems. They comprise several levels, each thought to obey to some central symmetry.[6] The biochemical, physiological, perceptual, and cognitive levels referred to in the introduction to this chapter are believed to exist in that increasing order,[7,8] although they may not necessarily correspond to the natural levels of the structural hierarchy. Additional levels, not yet identified, may also exist.

Living structure is maintained in an oscillatory equilibrium between two cyclic tendencies: a tendency toward symmetry, which is associated with the building up of new structure and order, and a coexistent tendency toward asymmetry, which we can associate with tearing down of structure and disorder. In biology, the first tendency is identified with anabolism and the second with catabolism. An organism, then, is a mixture of order and disorder, in perpetual alternation. This allows for a rather relaxed organization, maintained within a range of values rather than on a fixed set point. In unitary theory, the above tendencies are seen as complementary and cooperative, not antagonistic. Both are necessary to close the cycle.

Organisms are complex systems and their complexity prevents them from being isolated from their environment. Contrary to crystals, which reach a static symmetry and high structural order, organisms are only partially ordered and maintained in a process form in continuous balance with their environment. Although the organic hierarchies are self-sustainable, organic stability rests primarily on the stable patterns of some components, mainly proteins.

The enormous complexity of organisms, in their adult state and during its evolutionary history, may be more apparent than real. It is true that there are approximately 10^{29} protons and an equal number of electrons in each one of us, making up our atoms, molecules, cells, etc. However, there is only a very limited number of different atoms, and they are not dynamically independent. In spite of a gigantic number of fundamental particles, organisms display an extremely low number of degrees of freedom.

Concepts, such as that of function, can also contribute to the alleged complexity. The notion of function in contemporary biology was extrapolated from experience with machines. Each part of a machine is built according to a specific design, and its specific function is part of that design. In unitary theory, every specific function is evoked by some organic imbalance (generated asymmetry) and is thus part of the dynamics. The result of a cycle of function is the elimination of that imbalance (back to the same symmetry). The processes of structure-field-function are all aspects of one *single process*—the formative process going from asymmetry to symmetry. Ultimately, the function of a structure is simply to facilitate the repetition of the field process by which it was formed.[2]

In spite of a ceaseless interchange of individual atoms and molecules, the extensive coordination between the positions and orientations of the functional parts of an organism maintains itself. It is hoped that, when properly understood, this internal coordination will be one day described by just a few simple laws. These laws should then reveal the secrets contained in the "continuous normalizing process stabilized by hereditary units and an outer border", which was Whyte's working conception of animal organism.

Order from process

Order, like beauty, is a complex concept difficult to define in absolute terms. There is an element of subjectivity in its interpretation. The term also means different things in different branches of science. We are most interested here in *biological order,* the ordered arrangement underlying organic coordination. Since we do not know what this arrangement really is, no precise definition of order can be given. As a working concept, we may define biological order as a pattern of relations displaying some regularity or organization. The loss of this regularity is regarded as equivalent to disorder.

By thermodynamic theory, we have been led to believe that the general direction of physical events in the macroscopic world is towards an increase in *entropy*—the natural slide to disorder. This assertion, however, is against what is observed in cosmic and biological evolution, where order appears everywhere. Some assumption in the concept of entropy does not apply to physical and organic structure, thus preventing entropy from predicting the actual behavior of nature.

When theoretically analyzed, entropy is found to be a *statistical quantity* which is connected to disorder in an indirect way: in statistical thermodynamics, an increase in entropy is interpreted as a tendency towards states of higher probability; and in a collection of molecules, the more probable molecular arrangements are likely to be the less ordered, or more disordered, ones. So, an increase in entropy is identified with, and can serve as a measure of, *molecular disorder*. Entropy relates to probability, and probability, in turn, implies disorder.[9]

Entropy is also a very restricted concept of change. It has only been defined for systems divisible into parts, each of which is in equilibrium. As a consequence, it applies only to stationary states of systems, or to small deviations of these states, in which there is local equilibrium. Away from thermodynamic equilibrium, entropy loses its meaning. Most living systems work far away from equilibrium conditions and therefore lie outside the scope of entropy.[2,10]

These shortcomings, and their deeper theoretical implications, prevented the entropy principle from being used in the mathematical formalism of the unitary theory, the most general of all theories of science. A broader concept of change, directly related to local structure and capable of covering partially ordered as well as disordered systems, was required.

To the effect, entropy was generalized and transformed from a statistical (scalar) function into a new function explicitly dependent on local vectors of structural non-equivalences. This new entity represents the chiral deformation of a system from a state of stable equilibrium. It is a function of the spatial relations representing the relaxation processes of a system as a whole, passing from less ordered to more ordered states. Whyte named this entity *diminant*.[11]

As its name indicates, this variable continually decreases during a one-way process of decreasing asymmetry. The increase of entropy corresponds to the vanishing of diminants, so disorder in thermodynamics corresponds to the development of order in unitary theory. In this way, the tendency towards disorder or disorganization as a universal tendency of nature disappears from fundamental theory.

In the unitary conception, entropy is restricted to those processes in which the diminants represent differences of scalars. Thermodynamics is a one-way scalar theory while the unitary conception is a one-way vector theory covering processes of greater scope. It becomes evident that thermodynamics is a partial theory that ignores process and only process can be associated with the generation of order.

The diminant concept is based on the asymmetrical relations of space, time, and quantity, which are the ones also involved in every structural transformation in the organic realm, from organism-environment relations to those of growth, development and cell division. It appears, therefore, that these relations are the ones followed by nature.

In contrast to the view of thermodynamics, unitary theory proposes that structural asymmetry tends to slide naturally to symmetry, but is opposed of doing so by the operation of the field. Being always in process, the field constantly pumps asymmetry into cosmic and biological structure, allowing it to grow and to become more ordered and complex, thus negating to structure the tendency to slip into symmetry on its own. In this way, order is maintained and even tends to propagate itself. Where thermodynamics sees disorder, the unitary conception sees order instead.

But the concept of entropy and its connection to disorder still reigns strong in present physics and biology. To introduce the concept of order into living systems, Shrødinger coined the term *"negative entropy."* [12,13] This term obviously assumes entropy to be the primary property. To end the subservience of order and to emphasize its predominant role in life, Whyte proposed a new term for the processes associated with the development of order in organic systems: *morphic processes*. Entropic processes are those associated with destruction of structure. It is now clear that for structure to be destroyed it is necessary first to be built, so order is primary. It must therefore be measured directly in some way to be discovered, but not by way of disorder.

Morphic processes and entropic processes are complementary in action, and both are required for the maintenance of life. Entropic processes are related to catabolism, and morphic processes to anabolism. They should be viewed not as processes of order and disorder but as different aspects of the same process—the decrease of structural asymmetry in isolable processes. In effect, on a deeper analysis, it can be shown that both tend to equilibrium. Both are thus the result of the tendency to symmetry. Disorder is no more than the dispersion of order.

Structural order is generated during the development of form. In the course of the developmental process, past several finite thresholds, a characteristic pattern is finally achieved where the relative distances and orientations of its constituent atoms and molecules become stable. Once order is achieved, it tends to propagate itself; that is, once a stable pattern is produced, it tends to reproduce itself. The property whereby certain processes promote their own repetition is called *facilitation*.

We have already seen that stabilized form and therefore order can be achieved by identical multiplication of local specific structural patterns, such as those of genes and synthetic enzymes. These patterns retain within themselves a record of the process by which they were formed, which subsequently guides the process of their multiplication.

There are, however, less well known processes involving not local structures but extended stable systems of cyclically polarizable molecules, whose state of *orientational order* can be increased by intermolecular modifications introduced by a specific double pulse of depolarization-repolarization passing through the whole system. These modifications facilitate the repetition of later cycles, originated under the same circumstances. The development of functional structures—functions producing structures which stabilize the functions—and the processes of adaptation and of learning fall into this category.[2]

Facilitation, as conceived by unitary theory, is a very important property manifested not only by organic and mental processes but by inorganic ones as well. It oversees in what place order must occur, preferably here rather than there, and automates the response to frequent stimuli, from internal processes and from interactions with the environment. Without the process of facilitation there would be chaos. The property of facilitation, therefore, is at the basis of all order in nature and of all organization in human nature.[14]

The subject of order is as complex as life itself. We know only some separate aspects of the total order of an organism. Order, however, is everywhere: in cell membranes, inside and outside cells. The whole organism can be considered an ordered system working in close cooperation with a relatively disordered environment. Nevertheless, order is more evident in uniform regions made up of sets of identical proteins or other large molecules. Among those ordered regions are the semipermeable cell membranes and the conductive and contractile tissues. Their special molecular arrangement and physical properties make them akin to *liquid crystals*.

Liquid crystals are states of matter intermediate between the highly ordered solid crystals and the partially ordered or totally disordered liquids. They possess properties common to both of the extreme states but also have unique and quite remarkable properties of their own. Among those properties are fluidity, spontaneous orientational ordering, and exquisite responsiveness to a wide variety of stimuli—chemical, mechanical, and electromagnetic. These properties are found in all organic ordered regions.

Underlying all these properties are the shape and arrangement of the molecules in the materials that form liquid crystals. Characteristically, the molecules are oblong in shape (uniaxial symmetry) and arranged in arrays of parallel layers. In response to heat the molecules can take different positions but they all point to a given direction.

This special molecular behavior derives from the fact that the intermolecular forces in the crystalline solid are stronger in some directions than in others. When such a material is heated, the increased molecular motion overcomes the weaker forces but its molecules remain bound by the stronger forces. This produces a molecular arrangement that is random in some directions and regular in others. Positional order is lost, but full orientational order is maintained.

This mixture of stronger forces and weaker forces, of stability and instability, is found in the structure of all proteins at body temperature. The carbon atoms of a main-chain of a protein are lined up in an orderly arrangement, all connected by strong covalent bonds. These atoms have a repeated pattern, so the main-chain can be considered a solid crystal. On the contrary, the side-chains are held in place by weak hydrogen bonds and therefore they can afford some mobility. The molecular mobility of the side-chains is responsible for the properties of fluidity and quick responsiveness to appropriate stimuli which proteins share with liquid crystals.

But organic ordered regions are biological structures and so quite different from inorganic liquid crystals. Although both share the same normalizing tendency toward uniformity of polarization, only ordered regions, incorporated into living structure as components of more extended systems, can undergo repeated one-way pulses of depolarization and repolarization.

In each ordered region of identical protein molecules, each molecule is polarized in the same way, and therefore the directional properties of the resultant polarization cycle of each region as a whole correspond to those of its individual molecules. If some protein molecules are disturbed, they will tend to be reoriented, as well as other polarizable molecules such as those of the associated water clusters, so as to conform to the resultant field. All molecules of each ordered region form a working association with a characteristic functional cycle.

In addition to changes of net polarization, functional cycles also involve changes of shape in the individual proteins and cooperating molecules. The cyclic transformations of unitary theory thus correspond to the *phase transitions* of present biochemistry, referred to in chapter 3 and illustrated in Fig. 3 (pg. 53). In general, only changes of polarization are propagated but in certain ordered regions, such as those of contractile tissue, changes of shape are propagated as well. Muscle proteins are associated into myofilaments in such a way that molecular folding leads to macroscopic contractions, even against high resistance.

Of all ordered regions, the semi-permeable cell membranes play a very special role: their polarization state stabilizes all the polarized regions within cells and therefore within multicellular organisms. The polarization is brought about by the chemical difference between cell interior and exterior, which induces an asymmetry in each of the molecular dipoles, lipid bilayers and proteins, in the boundary. This collective polarization spreads not only along the membrane but also to the cell interior where, by imparting orientation and vitality to water-protein complexes, it sets the liquid crystal state of *cytoplasmic gel.* All functional cycles, however, can only continue as components of comprehensive one-way processes of relaxation and normalization passing through the whole organism.[2]

It is worthy of mention here, because of their importance as signaling pathways and energy transduction centers, some very localized ordered regions, made up of aggregates of atoms with an integrated vector, which are widespread in biopolymers, such as bone, collagen fibers, proteoglycans, and DNA. These highly ordered regions apparently result from the charge influence of asymmetrical carbon atoms on CO-NH peptide dipoles inside alpha-helices. They confer to the above structures *piezoelectric properties,* whereby electromagnetic oscillations are converted to mechanical vibrations in the structures themselves, and vice-versa.[13]

Organisms can thus be viewed as conglomerations of countless ordered regions, in perpetual change from asymmetry to symmetry and back again to asymmetry. However, for organisms to function as single structural units, all ordered regions and processes must be internally coordinated. This global coordination of the interior environment, which has escaped our understanding, must have been developed during evolution, along with the better-known Darwinian competitive adaptation to the external environment.[2]

In summary, life is maintained in a dynamical equilibrium between stability and instability, order being constantly generated and dispersed within limits imposed by an internal coordinative process which awaits clarification and, in a future quantitative unitary theory, mathematical expression. Nevertheless, whatever this global form of ordering turns out to be, it can only be sustained by perpetual interaction of organisms with their external environments.

Cooperativity and self-regulation

In a living system, processes do not occur in isolation. The processes of the constituent parts or units occur only in concert with those of the rest of the system. There is thus interplay among multi-processes. Each unit of structure or group of units is always in the process of reaching symmetry or stability of form, but this course of events is changed by the wider process of the whole system which continually tends to achieve overall symmetry as well.

The process of the whole is more stable than each of its unit processes and dominates them. Achievement of the symmetry of the whole is therefore the common aim of all system process units, and the tendency to reach that symmetry is the "force" that attracts the processes of the units into cooperation.

The development of the symmetrical form of the whole is brought about by modifications of the separate processes of their formative parts. Cooperativity in unitary theory is therefore a result of the formative interactions of process parts. In other words, the parts simply adjust themselves to a common developing symmetry. Within organisms, cooperativity of the processes of parts with that of whole is facilitated by the normalizing process.

The adjustment to the symmetry of the whole may involve changes not only in the spatial relations of the parts but in their internal processes as well. The intra-molecular changes, or transformations, always involve the passing of thresholds. As a consequence, the atomic pattern of the structure is altered, and so is its polar asymmetry. The unitary concept of *structural transformations*, intimately linked to that of cooperativity, is more profound than, and includes within itself, the concept of *phase transitions* of modern biochemistry.

In unitary theory, structure and field are inseparable entities and so the structure-associated field is also tending to symmetry. In the case of the field, this means tendency to become more uniform by reducing or eliminating field asymmetries or gradients. In order to do so, the normalizing field always tends to bring polarized or polarizable structures into those positions which conform most to the field. This auto-normalizing process of the field can excite into cooperation spatially separated neighboring structures, making them components of a single one-way normalizing pulse.

By rendering the field more uniform, cooperativity facilitates the normalizing of the field. Under these circumstances and within certain limits, every deviation from the normal polarization state (norm), which is the one with minimal polarization differences, is generally followed by its automatic restoration. The property of returning to the norm after every disturbance which is not excessive is called *self-regulation*.[2] These automatic adjustments, the result of cooperative processes, tend to maintain the structural stability. They allow every living system to regulate itself without outside intervention.

Since all systems tend to symmetry, cooperativity is found everywhere in the inorganic and organic realms. In animal organisms, cooperativity between structures and fields is typically found in all processes of synthesis and multiplication. It starts in the fertilized ovum and becomes increasingly efficient during differentiative development. All cellular processes are cooperative and all organic activity involves structural transformations or phase transitions.

We have already seen that the making and breaking of hydrogen bonds in water is a cooperative process: water clusters are constantly forming, dissipating, and forming again. In the unitary view, clusters can be seen either as tiny oriented aggregates with low thresholds or evanescent ordered regions, similar to liquid crystals, whose extension of order is denied by constant thermal agitation. An alternation of order and disorder, asymmetry and symmetry, can already be discerned in water.

Cooperative structural transformations appear to have some magic element of their own. In the case of water, when one hydrogen bond forms, many others will form, and when one bond breaks, typically a whole cluster will dissolve.[15] Thus, a change in the bounded state of a water molecule or a small group of molecules affects that of its neighbors. Unexpectedly, from this cooperative intermolecular binding and unbinding, a wave motion ultimately results.[16] There is no special agent at work, just cooperative interplay of polarizable molecules and fields.

In the case of solid structure, when organic gel systems of protein-ion-water cooperate among themselves, as components of one comprehensive normalizing pulse, other remarkable biological features appear to emerge: material transport, action potentials, secretion, multiplication and, in complex enzyme systems, oscillations.[13] And when animal protein is involved the most remarkable of all features makes its appearance—*motion*.

Gerald Pollack has described in detail, in words and in beautiful illustrations most of these cooperative processes.[17] He believes that "the phase transition phenomenon is the common denominator of cell function." But behind the transitional behavior of organic, many-particle cooperative systems is the elusive unitary process. By themselves, the phase transitions of structure would be unable to organize and develop. It is the normalizing process that induces them to do so, not only in a single cell but in all cells of the hierarchy, and adjusts them to the requirements of their environment. Otherwise, there could be no coordination, no temporal order, no growth, and no life.

Field asymmetry as life energy

For a theory to be general, it must be simple and profound. And, if this general theory purports to be a universal law of nature, it must select as empirical referent the basic field from which all structure derives. Unitary theory meets these requirements.

Energy is a measure of a system's ability to do work and its laws are the subject of thermodynamics. This branch of science is based on the concept of *internal energy*, which is a small fraction of the total energy of a whole system. The internal energy of a system corresponds to the kinetic energies of its constituent atoms and molecules and the potential energies associated with their mutual interactions. In unitary theory, the energy of all systems, from stars to organisms, emanates from the structured field, by way of the normalizing process. Energy is not divided into potential energy and kinetic energy. What exists is the unitary structured field with its content of energy.

The structured field is made up of three-dimensional asymmetrical particles, the distant precursors of atoms. Since energy cannot be distinguished from structure, it is conceived as structural asymmetry of any kind. This concept embodies the field with a geometrical shape. The field process then consists in the decrease of this asymmetry, and as the field changes form, energy is manifest.[8]

Instead of relying on the abstract concept of energy, unitary theory relies on the immediately observed *spatial-temporal relationships*. Isolable processes tend towards states, not of minimal potential energy but of higher spatial symmetry; that is, states with a greater number of symmetry elements. In unitary thought, it is not energy but spatiotemporal relations that are fundamental.

In thermodynamics there is no energy in process. This theory therefore can only describe isolated systems or, in the case of organisms, aged systems which have already settled down into cyclic or equilibrium states. One-way unitary theory is more powerful and capable of describing not only systems in isolation but the processes of isolation as well. Thus, it can adequately cover the processes of growth, decaying and aging that characterize living systems.

In the unitary conception, field energy or structural asymmetry is directly inducted into local molecular systems along a given axis in the molecule. This oriented process of induction is associated with a deformation of the structural pattern of the system: it is transformed into a more asymmetrical, polarized or activated state. The system is now ready to relax spontaneously to a more symmetrical state or, in the case of cycling structures, to its initial symmetrical state.

On cycling from asymmetry to symmetry, stable structures or organs remain unchanged, but in the cycling process some structural asymmetry (kinetic energy) is released and some work is done. This change corresponds to the anabolic phase of organisms whereby organic synthesis takes place to maintain or extend the organic pattern. To sustain the cycling process, structural asymmetry must be inducted again into organic structure from metabolic sources. This phase of the cycling corresponds to catabolism.

There are thus two separate but functionally integrated systems in living structure: one main functional system and an associated (coupled) supportive system, the supplier of polarization. In biochemical life, the polarizing agency is taken to be, for the most part, the ADP-ATP system.

It is an established fact in biology that the chemical energy of ATP is derived from the physical energy of sunlight by way of the photosynthetic process and the food chain. However, the unitary view of the process of energy conversion differs from that of biochemistry. It is based on spatial relations of structure as a whole, not on isolated chemical bonds. Energy conversions are mediated by structural transformations.

Specifically, the fundamental step in photosynthesis is the conversion of the point-centered symmetry of isolated carbon atoms into the translational symmetry of a carbohydrate molecule. The photonic energy is used to combine carbon atoms, derived from atmospheric carbon dioxide, with hydrogen atoms, derived from the photolysis of water, into chain or ring molecules possessing a high degree of symmetry and stability. These are the properties that make these molecules suitable as fuels for oxidation.[2]

The oxygen atom has special structural characteristics. It is missing electrons and is thus asymmetrical but its asymmetrical state is highly stable. It cannot easily be brought into a more symmetrical state. The ground state of molecular oxygen is also unique, in regards to the position and parallel spin state of its unpaired electrons, one in each atom.[13] This configuration embodies the oxygen molecule with a high capacity to contain the field.

The process of oxidation is interpreted as the induction of polarization by the highly asymmetrical, yet more stable, oxygen atom or by the activated oxygen molecule into the symmetrical but less stable fuel molecules. The induction of polarization or transfer of free energy from oxygen to the linear fuel molecules distorts them, and it is this distortion that allows the oxygen to rip off electrons from the fractured molecular structures to complete its own symmetry.

Oxidation is viewed, not as loss of electrons, as asserted by electrochemistry, but as induction of asymmetry in a stable linear molecule. Loss of electrons is not the primary process. It is the end result of the process of oxidation. This unitary concept of bioenergetics is schematically indicated in Fig. 7 (pg. 57), which connects the complementary processes of photosynthesis, where fuel molecules are initially formed, with oxidative phosphorylation, where final ATP synthesis takes place.

Regardless of whether energy derives from the nuclei of hydrogen, through the high temperature solar fusion reactions, or from the electronic ring of oxygen, through the low temperature organic oxidation reactions, it ultimately emanates from the basic field. In living systems, the non-nuclear energies of hydrogen and oxygen are intimately related. To access the "stored" energy of hydrogen, induction of structural asymmetry (free energy), either from oxygen or ATP, is required. The inducted energy is used to break the energy storage compartment, so to speak.

In addition to the sophisticated and efficient respiratory mitochondrial process, there are also local simpler respiratory systems, in cell membranes, vascular walls and elsewhere, consisting only of oxygen, fuel molecules and redox enzymes. In them, the structural asymmetry inducted into the fuel molecules by the oxygen field is directly transferred to the affiliated functional system. There is no formation of ATP. The functional system and its own respiratory system emerged during differentiative development and so there is a close functional fit between them. They operate as components of the same polarization pulse. This was Whyte's preferred view of metabolism.

In all earnestness, the light thrown by our present knowledge on the complex energy field is so dim that one cannot even see the players, let alone the strategy of the game. Whatever the latter will prove to be, it must involve the active participation or mediation of water (or the charged hydrated gel) in the energetic processes. It is believed, for instance, that the synthesis of ATP is powered by switches in water structure within the microcavity of the ATP synthase, occurring in synchrony with dissipation of the proton gradient.[18] And the fall in the structural energy of water clusters is thought to supply the energy for biological work.[19]

The high asymmetry of oxygen is not completely neutralized in water. It is only partially tamed by the two stabilizing electrons. A tug of war is continually going on between the oxygen atom on one hand and each separate hydrogen atom on the other, and the advantage always falls on the side of oxygen. It is from this uneven endless struggle that the permanent electrical dipole of a water molecule originates. Water dipoles and protein dipoles are packed together in the sketchy unitary theory of today, but they most certainly have delicate interactions of their own.[20]

It was energy, working through water, which initiated the organic realm. We can only imagine that with the increase in organic complexity this relationship would have become more sophisticated. If so, it would have reached its high point in the mental processes of the central nervous system.

SELECTED REFERENCES

1. GOULD SJ. The evolution of life on the earth. Scient. Amer. 271, 85-91, 1994.
2. WHYTE, LL. The unitary principle in physics and biology. The Cresset Press, London (1949).
3. LENTON TM, SCHELLNHUBER HJ. Climbing the co-evolution ladder. Nature 431, 913, 2004.
4. ROBERTSON DS. The rise in the atmospheric concentration of carbon dioxide and the effects on human health. Med. Hypoth. 56, 513-518, 2001.
5. MATTICK JS, GAGEN MJ. Accelerating networks. Science 307, 856-858, 2005.
6. WHYTE LL. Structural hierarchies: a challenging class of physical and biological problems. In: Hierarchical structures, edited by Whyte LL, Wilson AG, Wilson D. American Elsevier, New York (1969).
7. TABORSKY E. Evolution of consciousness. BioSystems 51, 153-168, 1999.
8. BARANSKI LJ. Scientific basis for world civilization. Unitary field theory. The Christopher Publishing House, Boston, USA (1960).
9. VON BAEYER HC. Disorderly conduct. The Sciences 37, 15-17, 1997.
10. WHYTE LL. One-way processes in physics and biophysics. Brit. J. Philo. Sci. 6, 107-125, 1955.
11. WHYTE LL. The atomic problem. A challenge to physicists and mathematicians. George Allen & Unwin Ltd., London (1961).
12. SCHRØDINGER E. What is life? The physical aspects of the living cell. Cambridge Univ. Press (1944).
13. CARVALHO JS. Life: its physics and dynamics. American Literary Press, Inc., Baltimore, MD (2003).
14. WHYTE LL. The next development in man. The Cresset Press, London (1944).
15. FRANK HS, WEN WY. Cooperative interaction in aqueous solutions. Disc. Faraday Soc. 24, 133-144, 1957.
16. WATTERSON JG. Does solvent structure underlie osmotic mechanisms? Phys. Chem. Liq. 16, 313-316, 1987.
17. POLLACK GH. Cells, gels and the engines of life. Ebner & sons, Seattle, WA, USA (2001).
18. WIGGINS PM. Role of water in some biological processes. Microbiol. Rev. 54, 432-449, 1990.
19. WATTERSON JG. What drives osmosis. J. Biol. Phys. 21, 1-9, 1995.
20. DEL GIUDICE E, PREPARATA G, VITIELLO G. Water as a free electric dipole laser. Phys. Rev. Lett. 61, 1085-1088, 1988.

CHAPTER 7

THE HUMAN NERVOUS SYSTEM

It is in the nervous system that the physical world meets the biological.
Sven Axelsson, 2001

After about one million years of simian evolution, *Homo sapiens* gradually made his appearance on the surface of the earth. The human species was endowed with a thinking organ of vast potentialities, well beyond those required for adaptation to the environment: the *brain-mind*. Over many generations, and under social and cultural influences, the hereditary characteristics of the human mind enabled the development of intellectual capabilities, expressed in the symbolic processes of thought. The intellectual supremacy of our species is at the basis of the dominant position we enjoy in the biosphere today.

At a superficial sight, the human mind may appear to be of a substance fundamentally different from that of matter. A deeper analysis, however, reveals that what makes mental processes so special is their time-oriented nature. Matter is neutral to time and indifferent to it, while the mind is "essentially oriented in time, being in some sense familiar with the past in memory, while anticipating the future in imagination."[1] This being so, only a theory of process can possibly explain mental processes.

Current biology is built on the quantum mechanical theory of local microscopic interactions that require measurements to generate information for the *a posteriori* conservation of energy.[2] The theory admits process but conservation implies permanence, and therefore this physical matrix is not appropriate for the description of the time-dependent mind. Instead, unitary theory builds biology on the physics of the asymmetrical relations—of space, time and quantity—and is thus capable of describing not only static systems but systems in constant spatiotemporal change as well. Being an explicit theory of process, unitary theory may be powerful enough to fully describe mental processes.

In the unitary interpretation, body and mind are both expressions of the organizing power of a single universal formative process, the unitary field process. Mental processes are thus formative processes which stand at the top of the complete hierarchy of the whole organism, from where they dominate and partially guide the material processes of the hierarchical levels below. In reality, the human individual is just one comprehensive formative process with differing degrees of complexity at each level of the organic hierarchy. Seen in this way, the terms material and mental, body and mind, or matter and spirit do not imply any general dualism; they are just two inseparable aspects of a single formative process.

The development of form is clearly manifested in mental processes, where separate patterns of relationships are intermingled, transformed, and extended in scope. When the brain is presented with a new stimulus, its form has to become part of the organized record of past experiences before the brain system develops the form of the new response. This cyclic autocatalytic process is repeated anytime a new stimulus is encountered.

The processes underlying mental responses are highly facilitated, but their great elaboration makes them very delayed when compared with the immediate instinctive responses or with the fast transmissive processes of the autonomic nerves. Time-delay is the quality that characterizes better the deliberate responses of the thoughtful humans. Animals inferior to man lack this characteristic but they can learn, and thus have intelligence, because the left over residues of their immediate responses are incorporated into the organized record of their memories. When confronted with the next similar situation, a modified pattern of instinctive behavior is revealed.

Mental and material processes both display a one-way causation but the inequality of their cause-effect relationship is different. In material processes a single stimulus is capable of generating multiple responses; on the contrary, in mental processes the cause is a multiplicity of varied stimuli while the result is a single idea or action. The one-way causation of mental processes proceeds from more complex causes to simple effects. This implies that the *invariant* in mental processes, the abstract persistent property that our intellect uses to correlate earlier to later events in a process, cannot be a physical quantity. It must be the expression of some unknown law of the development and transformation of patterns.[3,4]

The mental patterns, whose nervous activity contains the secrets of cognitive life and of all forms of human expression, are also different from the spatial patterns of structure described in previous chapters. Their intrinsic nature is not known for sure but, they were conceived by Whyte as patterns of energy configurations, extended in space-time and acting as a unit.[1]

From all quantum field cognitive theories, unitary theory is unquestionably the most powerful and the one capable of given us some concrete idea of how the coordinative structure and mode of operation of psychological systems may be accomplished. But many unknowns still remain and a comprehensive scientific theory of the mind is not yet at hand. Furthermore, unitary theory still awaits confirmation, but all these facts should not prevent us from giving a general unitary description of the central aspects of mind function.

The unitary view of brain organization and functioning

In his proposal of the unitary theory, Whyte envisaged the basic field as a structured entity composed of sub-atomic three-dimensional structures to be identified. Induction of free energy into a system was taken as induction of structural asymmetry.[3] Later, Baranski postulated that the primary field particle was a three-dimensional free-energy structure from which all known structures derived. He regarded the induction of free energy into a system as the induction of these free-energy structures. It is on this extended field concept that modern psychological unitary theory is based.[5]

According to Baranski's theory, all atomic structure is conceived to be a composite of quantum field free-energy structures, maintained more or less stable by the nuclear and electromagnetic forces. The atomic-molecular structure, including nucleoprotein organizations, is further bound together by strong interatomic Heitler-London bonds, which keep molecules in shape and by lattice van der Waals' forces. These bonding constraints do not confer enough plasticity to biomolecular structure to allow it to function in genetic and psychological processes.

It is therefore postulated that the atomic nuclei of the genetic and mental memory structures must contain, somewhere in their framework, highly dynamical and flexible quantum field structures or special aggregates of these structures which are the *basic units* of memory processes. Under this view, the required atomic-molecular stability is provided by nucleoproteins, nuclear DNA for the genetic memories and, presumably, axoplasmic neural RNA for the psychological memories.[5]

It is a fact of nature that the one-way development of systems, physical or biological, cannot pursue by a monotonic process of increasing complexity in the system itself. Beyond a critical degree of complexity, a system loses regularity and its behavior becomes unpredictable and chaotic.[6] To sustain regularity, evolution has to arrive at the formation of a superimposed level of structure in the system. After that, development starts all over again, in a more highly organized and efficient way.

As a result of this natural principle, our nervous system ended up with three hierarchical levels of organization, which were developed slowly during the evolutionary process of extension and specialization of the spinal cord: the tropistic or physiological level, the perceptual level, and the more recent symbolic-conceptual level. This simple hierarchical architecture probably represents the prime mode of coordination or organization of mental processes and of their expression in the symbolism of language.

The increasing level complexity refers to the degree of organization and number of memory centers involved: one on the tropistic, two on the perceptual, and three on the cognitive level. Besides the memory centers, which are composed of separate but interrelated sensory and motor areas, each hierarchical level contains a central structuring center where the intrinsic formative tendencies of the incoming quantum stimuli are allowed to occur. It is from the processes taking place in these structuring or formative centers, synchronized by the normalizing field, that all our psychological phenomena are generated.

An environmental or external stimulus is viewed as a varying structural pattern of electromagnetic, mechanical, thermal or chemical energy. There are two different variables in a stimulus pattern—its configuration or form and its asymmetry level or free energy. Using the language of classical physics and information theory, we might say that a stimulus is "informed energy." This macro-mosaic type of pattern is transformed, within the nuclei of the neural receptor cells of the respective sense organ, into temporal series of translationally-moving quantum field structures, each one carrying a part of the structural configuration of the original pattern.

On moving in and out of the nuclei of the external receptor cells, the quantum field structures trigger electrical changes in the whole atomic-molecular receptor systems, which ultimately lead to the generation of the nerve impulses. The latter subserve mainly biochemical processes, while the quantum field structures are the basic units of psychological processes.

In general terms, all external receptors of the major sense organs (vision, hearing, balance, taste, smell) and skin receptors (mechano, thermo, and pain receptors), as well as all internal receptors, feed directly into the first level of structuring. The internal receptors comprise the externally-oriented proprioceptors of skeletal muscles, tendons and joints, and the whole variety of the internally-oriented smooth muscle receptors of organs and glands. This sensory orchestra then sweeps through all memory centers of the three structuring levels until it reaches the third and highest cognitive memory area. Presumably, before crossing the memory areas of a given hierarchy, the sequence of patterns constituting a stimulus is momentarily dispersed in space so that each component pattern can decouple and carry with it, to the structuring center of the hierarchy, any configurationally similar memory pattern previously recorded.

A recorded memory is a stable quantum structured organization composed of a central pattern and several side chains. The central pattern is the particular type of quantum memory structure that the memory area is differentiated to subserve; the lateral chains are the quantum memory structures of all external and internal modalities that were temporarily contiguous with the central memory at the time it was recorded. These side chains are also carried along when the central chain is decoupled. The decoupled quantum memory pattern is almost immediately replaced by the local synthesis of an identical replica, a spatially directed process of induction facilitated by the permanent "record" embedded in every memory structure.

Before moving from one structuring level to the next, all these quantum memory organizations undergo complex structural transformations in their respective structuring centers. Memory patterns of a given modality are first individually transformed and then combined into a unified whole (psychological whole) manifesting qualitative properties absent from its components. These structural transformations (*transactions* is the appropriate unitary term) are carried out by the unitary field process, and they are the source of all our perceptions, emotions, concepts, thoughts and actions.

All psychological processes, as their physical and biological counterparts, are brought about by the two tendencies of the unitary field working in close cooperation: a first tendency toward symmetry and a second tendency toward uniformity of asymmetry in the field as a whole. In biological systems requiring extra amounts of energy for their activities, the second field tendency works through the normalizing-respiratory process. In brain structure, these two tendencies are postulated to operate in processes that temporarily become isolable in specialized quantum organizations within the reticular formations of the brain stem—the structuring centers. These reticular centers are conceived to contain the facilitating molecular organizations for the dynamic formative process to take place. The tropistic structuring center, to which all sensory quantum field structures converge, is located in the thalamus and the other two centers in the higher brain stem.

On arrival to the tropistic structuring center, each dispersed quantum field structure of a given stimulus, together with each decoupled quantum structure of the same modality, all in differing degrees of symmetry and asymmetry, are allowed to follow their intrinsic symmetry tendencies and to coalesce into *quantum structural aggregates*. In this aggregating (transacting) process, the discrete quantum structures are brought back into spatial relations equivalent to those in their original stimuli.

After the stimuli from each sensory modality active at the time have been spatially structured, each sensory modality feeds its structural aggregates into a contiguous central collecting structural organization where they are immediately inducted with structural asymmetry and coupled in parallel, one at a time, onto the common asymmetry chain of the normalizing process passing through the region. It is during these brief moments of symmetrization, asymmetry induction, and serial coupling that conscious psychological phenomena are postulated to be experienced. The tropistic sensation forms the "ground" of our sensory and affective perception, but in the human brain it is not felt in isolation. It is synchronized with that of the other two levels above.

Before being carried away to the next higher hierarchical level by the normalizing field, each quantum structural aggregate is replicated twice, by field induction. One copy feeds back to the sensory memory areas that were just crossed, where it extends and differentiates configurationally similar memory structures, which then become part of future stimuli. This memory differentiation process, facilitated by the quantum memory structures themselves, appears to be quite slow. As a consequence, the feedback activity persists for some time after the stimulus has ended, leading to a "to and fro" movement from structuring centers to sensory memory centers and back. This is more pronounced at perceptual and cognitive levels where most of the learning process takes place. It is also during this prolonged feedback activity that the proprioceptive stimuli from internal motor receptors are fed into all other sensory modalities.

The other copy is sent to the complexing memory centers of the external (striated) and internal (smooth) muscular systems, on the same hierarchical level, for specific muscular contractions. Dispersion of the structural aggregates, not into discrete quantum structures but into fragments, must also occur followed by decoupling of past similar memory chains which are carried to special centers where the sequences of external and internal muscular contractions are controlled. The external motor control center is thought to be located in the corpus striatum and the internal control center in the hypothalamic supraoptic and paraventricular nuclei, where connections with the autonomic nervous system and the endocrine system are established.

At the top of each hierarchical level, there are thus three possible pathways for a recently formed complex quantum structural aggregate, or its identical copies, to follow: forward to next higher level, forward to motor areas, and backward to sensory areas. But when the highest hierarchical level is reached, these pathways become only two. Here, the cognitive sensory feedback pathway acquires a special importance for it is capable of continuously connect and pace all structuring levels below.

For simplicity a recorded mental memory has been treated as a single structural entity. However, it is thought to be a two-structure composite, similar to its genetic counterpart at biochemical level. Each structure or chain of the pair is located in each of the two strands of the neural RNA nucleoprotein. The chains are homologous and in continuous interaction, but only one allows quantum field structures to be decoupled from it. The decoupable chain is involved in the initial complexing of stimuli and the undecoupable is involved in the feedback memory forming process.

The number of memory areas and the degree of differentiation of their quantum organizations increase from tropistic to cognitive levels. As a consequence of this hierarchical distribution of complexity, most of the quantum structures supplied to the tropistic and perceptual structuring centers derive from the environment, while those supplied to the cognitive center derive largely from the organism.

At cognitive level, the degree of structural complexity is enormous. To sustain this high degree of complexity, the brain structure must be endowed with an equally high degree of functional order. Most of this order results from the selective mode of action of the structuring process: only one structural aggregate can be spatially symmetrized at one time. Each aggregate, or chains of aggregates, is then inducted with structural asymmetry and laterally coupled onto the fast-moving common asymmetry chain. Thus, temporal order follows spatial order.

This orderly coupling onto the asymmetry chain brings structural aggregates of the same modality into spatiotemporal interrelation and provides for interactive exchange between the various sensory and motor modalities. It also provides for the temporal relation between the sequential events taking place in the outer world (the exteroceptive stimuli) and those taking place inside the organism (the interoceptive stimuli). The structural aggregates, the units of order, finally reach the memory areas, where they are recorded and from where they can be recalled and used to order present or future thought and behavior.

The particular nature, complexity, and momentary variations of our experiences are determined by the particular momentary nature of external and internal stimuli and by the nature of the complexing quantum memory organizations they carry. In humans, the tropistic and first perceptual memory areas are thought to contain mainly short and relatively fixed chains which were philogenetically acquired. Most of the ontogenetically acquired memories are stored in the second perceptual memory centers located in the frontal lobes. Their long and complex chains are constantly updated by the rotating feedback loop emanated from the perceptual structuring center, a process going on throughout the lifetime of an individual.

The so-called "instincts" are complex and relatively well-organized quantum memory structures already present in the organism (second perceptual and first cognitive memory areas) at the time of maturation. These quantum structures, when decoupled by incoming stimuli, tend to promote the relationship of organism to environment. If they exist in humans, they have been superseded by more complex, efficient, and flexible forms of activity. They are now controlled by the cognitive processes.

But, how can we actually perceive things? And what processes, neural or otherwise, allow us to do that? According to unitary theory, the universal formative process consists in the development of form by the decrease of asymmetry. Since our psychological phenomena are resultants of asymmetry to symmetry changes in the structural aggregates, we experience directly the unitary field in operation. It is the intrinsic process of dynamical aggregate formation that yields the figurational (form) and dimensional (intensity) aspects of our perception. The elements of the stimuli of each sensory modality coalesce into a single aggregate in a very specific order, and during this continuous but constantly varying structuring process we experience definite forms, with all their qualities, intensities and variations. We are able to observe these forms because of the hierarchical nature of our nervous system; that is, our higher and dominant cognitive processes are capable of scrutinizing events occurring at a lower level.

In the formative process of perception, what is coming into the system (the stimuli) is correlated with what is already present and ordered in the system (the past historical memory). The historical memory was acquired slowly during the evolutionary process of our senses. Thus, behind *form perception* there is a high degree of complexity and differentiation of the quantum memory structures of whatever modality. Taking vision as an example of external perception, the quantum field patterns decoupled from the first perceptual memory area by the visual stimuli, when structured, yield our sensation of colors (hue variations), and the more complex configurational patterns decoupled from the second memory area yield our perception of objects, situations and events in space. Similarly, in hearing, the first memory patterns yield the sensation of sounds (ranges in pitch) and the patterns of the second memory yield words and melodies.

Historically, the salient steps in the evolutionary development of the above sensory modalities, from their physical beginning to their psychological end, appear to have been as follows: a tiny fraction of the electromagnetic radiation, called light, was first perceived as color, and objects were later perceived by the color of their surfaces. Some less-energetic electromagnetic radiation, the auditory part of the spectrum, was first perceived as sound, and words were later perceived by the sounds they produced. The dimensional aspects of perception, such as the brightness of light or the loudness of sound, are conceived to be brought forth at the moment of asymmetry induction and parallel coupling onto the common asymmetry chain. The intensity of a stimulus is a direct function of its original asymmetry level.

Without knowledge of the unitary principle, it would not be possible to understand the origin of our affections—the rise of our emotions and the quality of our feelings. For centuries, this so-called "hard problem" of psychology has been defying rational explanation. The nature of the "felt" part of the affective phenomena was thought to be different from that of the brain structure from where it originates. But in the unitary view it is also a momentary manifestation of the structured field in the brief moments of the formative process taking place in the reticular structuring centers. In this case, however, the affective stimuli, the carriers of the specific quantum field structures responsible for the phenomena, come not from the external environment but from the internally oriented smooth muscle receptors, particularly those of the visceral organs of the gastrointestinal tract, heart, glands and small blood vessels. Together with the decoupled quantum memory organizations of the same modality, these multiple stimuli, when spatially symmetrized and energized in the reticular structuring centers, yield our diverse affects. At perceptual level they are experienced as basic emotions—fear, anger, pleasure—and at cognitive level as the more refined affects known as *feelings*. The building of our personal character depends in large part upon the proper development of these inner feelings.

The stimuli from the skeletal muscular system are perceived as kinesthetic impressions of specific muscular movements, which are felt in a more refined way at cognitive level. These internal motor stimuli originate in tiny receptors densely packed in our muscles, tendons and joints, known as proprioceptors. These "body sense" receptors precisely measure physical properties, such as muscle length, tendon tension, joint angle, and deep pressure.[7] The trains of quantum field structures that constitute their stimuli, together with those supplied by the touch receptors, are fed into the various sensory modalities through the to-and-fro feedback pathways. Among these modalities are those supplied by the vestibular organs of the inner ear and by the visual system. Complex quantum structural aggregates then carry all this patterned information to the respective motor areas where appropriate muscular contractions are evoked.

The tropistic memory area contains quantum structures that can only generate uncoordinated movements of turning and bending, as those manifested by a child in the first six months of life. With numerous repetitions of particular movements, the memory organizations become increasingly differentiated in the sensory and motor areas, and with time the muscle control center becomes capable of achieving muscular coordination. However, it is at cognitive level, and after many years of practice, that the more sophisticated muscular contractions, such as those involved in language, sports, dancing, and in the use of tools are finally accomplished.

Perception and cognition are not different phenomena. Both are manifestations of formative processes, occurring on different levels of complexity. The differences in properties reflect differences in the structural configurations of their quantum structures. At perceptual level, the structural aggregates yield percepts and at cognitive level they yield concepts or personal attributes. They are thus interrelated and interdependent processes. Their ultimate aim is to adapt the organism to its environment.

This adaptation is carried out by purposeful muscular contractions and by control of posture. Under the influence of the somatic nervous system, the externally oriented skeletal muscles adapt the organism, through locomotion and manipulation, directly to the external environment; and under the influence of the autonomic nervous and endocrine systems, the internally oriented smooth muscles control the internal balance—that is, they maintain the fluid matrix in a steady-state. Since hormones, through the release of enzymes, are the controllers of metabolism, the internal muscular system supplies the structural asymmetry required by the external muscular system. These two systems are thus in intimate interaction, and their balanced coordination is sustained by the comprehensive normalizing process of which they are components.

The thinking process and its states

All mammals interpret and respond to their perceptions in a discriminative way, but only humans are capable of reacting to their environment deliberately or swiftly, as the occasion demands. This is a reflection of the high degree of differentiation and organization of their central nervous system. By allowing for great specificity of response, deliberate cognitive processes, together with other special faculties, have enabled humans to control their environment and their own fate.

What took to be human was the genetic acquisition by the species of the third cognitive memory area, which specialized in the recording of complex quantum structural patterns in abstract form. This memory specialization allowed the perceived environment to be interpreted in terms of long, complex structural configurations, which are required for conceptual-symbolic processes, such as those of language, to be manifested. The extension of the cognitive hierarchy was followed by further differentiation of its memory organizations, and the perceptual and motor processes became increasingly more discriminative.

But the conceptual-symbolic processes of humans are not developed at birth. Babies can cry but cannot talk. Humiliating as it may be for the dominant species, human babies cannot even walk. They must *learn* how to walk and talk, and how to think. Contrary to genetic memories, which are transmitted from parents to offspring, the majority of perceptual and cognitive memories must be assembled by each individual, the best way he or she possibly can. We learn by the ontogenetic formation of sequences of quantum patterns, a process which starts in the very early of our lives.

In the adult, the first cognitive memory area contains the complete records of past experiences, with all attendant aspects and affective and motor components; on the second cognitive memory level are recorded generalized past experiences called *concepts*, the counterparts of percepts, at the perceptual level. These generalized past experiences are contained in long, complex side-chains of similar quantum structures (sensory, perceptual, and cognitive) which yield, upon structuring, ideas, attitudes, interests, values, etc. Finally, on the third and highest memory level, characteristic of the human species, are the records of generalized concepts with language symbols. Each one of these records seems also to contain long and very complex sequences of quantum structural organizations, but it is recorded in a compacted, closely interrelated pattern. These records are called *abstract concepts* and *personality traits*.

Abstract concepts are as real as concrete concepts. They simply contain the maximum of meaning, generality and representation in a very small and highly complex structure; similarly, each personality trait has the essentials of all lower areas within its structure. Preservation of flexibility in face of extreme complexity may have dictated the abstraction process. Abstraction then permitted the reproduction of symbolic concepts and personality attributes in human conceptual-symbolic thinking.

The cognitive level of structuring is at the top of the total hierarchy. When the aggregate-coupled common asymmetry chain emerges from the perceptual level of structuring, it can pass either through one or all three complexing memory areas on its way to the cognitive structuring center. After structuring is complete, the nascent cognitive structural aggregates have only two pathways to follow: forward to the cognitive motor memory areas and backward to the sensory memory areas that were just crossed. It is postulated that this backward pathway continuously feeds back into the cognitive memory areas and back again to the central structuring center. This continuous loop of structural activity, called *central process,* is the source of our abstract-conceptual thinking.

Apparently, the two highest cognitive memory areas undergo continuous organization, elaboration and extension throughout the lifetime of an individual. This cycle of activity goes on constantly, unless perceptual processes disrupt it by evoking the *attention process.* When this occurs, the central process synchronizes cognitive with perceptual processes making possible the selective induction of configurational asymmetry into the highest perceptual memory area. This perceptual-cognitive synchronization is capable of giving cognitive meaning to our perceptions. It is at these moments that we are able to interpret, judge, and reasoning the significance of our experiences.

Basically, the central process is the normalizing process in a differentiated form, working at the highest level of the hierarchy. It enables the human to order and reorder his conceptual memory sequences and to form new sequences that further extend and differentiate the memory organizations already present. It is from this structuring activity that the attributes of the personality, its goals, purposes, attitudes, beliefs, interests, and values, ultimately spring. Since the central process can send quantum memory organizations to the motor areas, any of these personality attributes can, at one time or another, motivate our behavior. Thus, through the central process the human can precisely control his thinking and behavior.

The central process is the source of our imagination, our creative thinking. But in so doing the continuous central process does not always work at a high activity level. Most of the time it works under threshold level as a continual organizing and ordering process. This state of dim or low asymmetry level corresponds to the old notion of *"unconscious".* It is in this subthreshold state that our ideas are incubated and where our imagination works. The "unconscious" state is also the inner place of contemplation and the site of the "automatic pilot".

If anyone is confronted with a specific problem to solve or some cloudy past events requiring clarification, leave them to the subthreshold central process. You do not need to read more, to think more or to do anything else. Just bear the problem in mind. The digestive or associative process may require days, weeks, or even several years of incubation, but a solution or a more logical explanation is always found. In this regard, the "unconscious' state is a hidden source of power.

Many people use this creative thought process secretly or automatically, but everybody should know about it specifically, particularly the young, and be actively encouraged to practice it. Spencer

Brown tells us that Isaac Newton practiced contemplation routinely,[8] Baranski reveals that the physiologist and physicist Hermann Helmholtz "often got his clearest insights on complex problems while walking in the garden,"[5] and Whyte suggests that we should explore the contents of our unconscious at least once a day, in the morning, to "collect the harvest," as he puts it.[9] Obviously, we should always critically evaluate our unconscious insights before accepting them.

The structural meaning and nature of consciousness

The unitary mental theory of Baranski, which we have been exposing in a very succinct way, if confirmed, will undoubtedly be a gigantic step forward in the understanding of our thinking organ. A great deal is still unknown, to be sure, but the theory has already given us a new and long-desired interpretation of the unconscious, separating it from another mysterious entity known as *consciousness.*

The term consciousness has a number of different connotations, ranging from phenomenal awareness of one's perceptions and sensations to self-awareness, the perception of oneself as an agent endowed with intentionality and free will. But the major quest throughout history has been how consciousness arises from physical processes in the brain.

In the unitary interpretation, psychological processes are due to changes taking place in the structural aggregates of quantum field particles within the reticular centers. Consciousness, however, does not result from structural transformations. It is due to the inner symmetry tendencies of the numerous quantum field structures that comprise the asymmetry chain passing through the reticular formations.[5]

Consciousness ceases to be that mysterious entity referred to as "mind" by the philosopher. In effect, it is no more than a structured field in a high degree of symmetry tendency. Saying it another way, high concentration of quantum field structures and high rate of asymmetry to symmetry tendency yield the phenomenon of consciousness. The "stream of consciousness" is provided by the continuous and rapid rate of coupling of the structural aggregates of the various modalities onto the fast-moving asymmetry chain. Several degrees of awareness or consciousness may occur according to the number and complexity of the structural patterns and the amount of structural asymmetry flowing in the asymmetry chain. For this reason, there is no single condition called human consciousness.

As long as the essence of mental and material processes was considered to be of different origin, our comprehension of psychological phenomena and of consciousness wouldn't have been possible. But nature, body, and mind must be seen for what they really are: components of a single process. And a process cannot be broken. If it is, a dualism then arises. Virtual world and real world, subject and object, body and mind, are all dualisms and as such no longer tenable.

The continuity of the structured field from physical to biological to mental structure does not require conversion of bosons into electrons or energy into information to cross imaginary cuts in energy flow. In unitary theory, free energy and information are both contained in the patterns of structure, and they are already present in the basic field.

Historical development of the thinking process

Thought development is not an arbitrary process. It is the result of progressive differentiation of memory organizations at cognitive level, under the guidance of the normalizing field. This differentiation was initiated by a new kind of changing environment, the socio-cultural environment. Slowly, over a long stretch of time, the differentiated memories led from simple and fast to complex and delayed forms of thought and behavior. The development of thought is part of the one-way development of the human personality and of adaptation to the environment.

Although evolution of the human cognitive processes have been under way for the last 200,000 years, it is only over the last 10,000 years or so that the development of the thinking process has been accelerating. Prior to that, intellectual development was slow. Primitive man of the Stone Age had few conceptions, a short memory, and limited faculty of speech. He did not know himself as separated from nature. His behavior was spontaneous and unchallenged, guided only by the fast acting perceptual-instinctual processes. Some exceptional men and women may have lived in this period but their influence on the tradition was not durable. Communication was only by example and mimicry. Culture was based on superstition but, later in this period, some vague relationship to a superhuman deity, an undifferentiated concept of soul, may have been already brewing.[5]

This slow and harmonious way of living changed dramatically in the period from about 8000 to 1000 B.C. The *primum movens* for the change was the development of agriculture and stockbreeding, which made urban living possible and led to the emergence of a new environment, the socio-cultural environment. With community life, interpersonal relationships became closer and romantic adventure easier. The spoken word grew in power and, later in this period, the written word made its appearance. As a result of these changes, the mental processes were developing faster than ever before.

Intellectually, however, there was still no separation of idea from fact. Man was already thinking but he continued to be undifferentiated from nature. He was still an extrovert pagan guided by his instincts. With more leisure time available and higher technical skills, the societies of the time became involved in constructive activities. These were the first steps toward the beginning of modern culture and the building up of the early civilizations. These activities were particularly intense in the Eastern Mediterranean region and in Asia.

Although the alimentary conditions improved for all, the profits of the rising economy increasingly became the privilege of just a few, and from this economic unbalance a social hierarchy began to emerge. At the same time, romantic adventure became the sole prerogative of the rich and powerful. This unfair situation, unequal distribution of the economic surplus and denial of the most desired sexual pleasures to the vast majority, led to a strong emotional reaction and feelings of jealousy, frustration, and anger, and willingness to fight at the slightest provocation became widespread.

By 2000 B.C. some communities were degenerating and something had to be done to avoid chaos. With few options available to stop the violence and calm down the emotions, the religions and philosophical systems of the time gradually developed some rules of conduct, a system of taboos and mores, which we may call morals. These repressive rules, which associated feelings of guilt, fear, shame, and sin to the pursuit of the instinctive sexual pleasures and economic profits, were passed on in the cultural memory of the communities.

With the passage of time, the concept of organizing behavior to fit the changing social conditions became more defined. The attention of the individual was increasingly drawn to his own thought, as well as to the external stimuli, and he became aware of himself as thinking and feeling person with the faculty of choice. The cognitive central process had reached enough differentiation for abstract thinking to take place. By 1000 B.C. the new rules were challenging the dominance of the earlier perceptual-emotional response.

It was the beginning of the triumph of religion over instinct, not over a properly coordinated instinctive life but over an exploitation of instinctive pleasures that were damaging to the society. But instead of inner harmony, a tug of war was being played between thalamus and nature on one side and cerebral cortex and the new ideology on the other.

After 1000 B.C., the developing intellect was generating concepts of increasing generality, and ultimately universal ideas were conceived, the most important being the concept of one universal god. The origin of this great and noble idea is obscure, but it may have resulted from an elaboration, by the Egyptian priesthood, of the old concept of soul.[5] Monotheism came to prominence in this period, replacing the earlier polytheistic religions. However, its new ethical ideas (god, spirit, truth, faith, obedience, goodness, etc.) continued to organize behavior by the use of static concepts, while spontaneous behavior continued to express a formative process. In this way, a disjunction or dissociation grew up between the organization of thought and the organization of nature.[4] The old morals and ethics were somehow taken to mean that the pursuit of economic gain and romantic adventure were not socially desirable in themselves. The organic continuity of nature-perception-thought-action was not properly integrated. The tradition was in the course of developing a system of ideas for organizing deliberate behavior which had no direct relation to man's instinctive and spontaneous activities.

Like a confused adolescent, the individual was torn inside by feelings he could not understand. He was aware of himself as a thinking person separated from nature, but his thoughts were pushing him more and more away from nature. His control over nature was diminishing and unconsciously giving him a sense of anguish. At the time, the authority that man could most easily conceive was that of a father or person, so the situation was solved by developing the concept of a personal god. God was treated as the mediator between man and nature.[4] This new integrated principle required behavior to be in conformity with god's wishes.

These were the times of the great empires (Egyptian, Persian, and Chinese) and of intense trade and commerce, particularly in Europe. The techniques of travel and communication were improving and the new ideas were spreading rapidly throughout these regions. Philosophy also sprang up in this period when the Greeks, discontent with the supernatural and mythical explanations of reality, began to suspect that there was a rational or logical order to the universe. All of this was occurring centuries before exact science came into being in 1600 A.D.

Among the early Greek philosophers, Heraclitus (540-475 B.C.) stands out as the first to formulate the concept of process, in contrast to the prevailing static ideas of the time. Most of his surviving work reached us indirectly through the writings of Plato (427-347 B.C.) and in the form of epigrammatic sentences. An example: "Heraclitus says somewhere that all things are in movement and nothing stays put, and likening the real to the flowing of a river he says that one could not step twice into the same river."[10] The river, like the universe, is an example of unity in constant change. The unity of nature was to be found in its variety, and in the continuity of the transformation by which life became death, and night day.[4]

For Heraclitus, harmony was not to be regarded as static but as lying in a developing relation, or interplay, between opposites. "Disease makes health pleasant and good, hunger satiety, weariness rest." Change was universal but there was a pervasive order within it. Man was a part of that order and subject to its transformations. As a man, Heraclitus was still undifferentiated from nature. His ideas were too advanced for the time and, lacking scientific support, were difficult to grasp by minds molded by static concepts. It would take 2,000 years for the vague Heraclitean conception of a natural formative process to be definitely established.

It should be noted that the concept of change of Heraclitus was different from that proposed by the atomists, the most well known being Democritus (460-370 B.C.). These philosophers believed that the universe consisted of empty space and an infinite number of indivisible and eternal atoms. Change in this theory is simply the relative motion of permanent entities, which do not themselves change.

Plato accepted the validity of the world of the senses of Heraclitus but rejected the phenomenal world of process in favor of a transcendental world of permanent ideas. He did not understand how we could obtain knowledge about the permanent and non-changing properties of all that is around us when the information reaching us is changing all the time and when we ourselves are subject to change.[11] Obviously, he could not be aware that our nuclear memory organizations, from where our knowledge derives, are stable structures.

Where Heraclitus saw harmony, Plato saw discord and confusion. Reality was to be found in the persistent harmony of the ideal world. Platonic thinking was in tune with the subjective religious ideas of the time and persisted, while the process idea of Heraclitus, like an unknowingly beautiful flower unable to blossom in a hostile environment, withered and remain dormant. And a permanent dualism between the process of the senses and the permanence of timeless ideas became established.

Religion and philosophy dominated the intellectual world up until 1600 A.D., when exact science came about. Throughout this long period, the conscious mind was king. The "state of consciousness" was endowed with a special metaphysical status and taken to be the supreme control of behavior. In retrospect, only a strong religious influence on philosophical thinking could have led a philosopher to ascribe such powers to simple transient moments of attention to particular forms. With no integrating concepts available, the dualistic religious and philosophical ideas were the structures developing the mental processes and sustaining the moral systems.

In the West, this time period corresponds to the Dark Ages, which were characterized by economical, social, and cultural stagnation. In the East, a similar course of events occurred, but early on in this period there were notable scientific accomplishments, the most significant being those of al-Khwarizmi (780-850) in algebra and Ibn-al-Haytham (965-1040) in optics, which greatly influenced Kepler and Descartes. It took, however, over two centuries for these works to be translated in Latin and become available in Europe. This pace of transmission of knowledge changed abruptly when the printing press, also imported from the East, was introduced in Europe in the fifteenth century.

About 1600 Kepler (1571-1630) and Galileo (1564-1642), working independently, formulated the principle that the laws of nature were to be discovered by measurement. The structure of nature was wholly quantitative. What was not susceptible to measurement, it was out of the scope of science. By emphasizing the object, this way of thinking encouraged the decline of confidence in the powers of the mind. Man was considered as a machine. In science, this principle became the basis of the scientific method; in the social field, it became a technique of expansion. Since all magnitudes have equal status before the laws of arithmetic, expansion has no limits in any field, be it in wealth, power, territory or even family. There is thus an intrinsic anarchic aspect in the concept of quantity.

Over the ensuing centuries, the quantitative principle led to great achievements, in theoretical and applied science and contributed to a decrease in poverty and disease. Out of Galileo's first principles of mechanics, for instance, came the quantum mechanics of the twentieth century. The prestige of quantity was enormous and has extended to our days. The American ideal of personal achievement is still based on it. But there is no room for form, order or organization in the concept of quantity. Quantity represents static permanence and the concept cannot express the forms of mental processes.

The supreme role of religious self-consciousness was being challenged by the quantitative principle; the subject was yielding its supremacy to the object. Objective science was replacing subjective religion. Descartes (1596-1650), however, was still maintaining that the truth could only be found in the deliberate processes of the conscious mind. He believed, wrongly, that consciousness was a primary, substantive reality. But it was an intellectual, not a religious entity. With Descartes the dualism of mind and matter was clearly established. The development of form in nature and the origin of ideas were not understood. There was still no return to a process concept. The European tradition, therefore, was still growing inadequately.

Mathematics flourished in the seventeenth century. Descartes introduced analytical geometry and Newton (1642-1727) the differential and integral calculus. The latter techniques allowed Newton to formulate his laws of motion and gravitation that dominated the scientific thinking of the eighteenth century. Reduced to cases of matter in motion, all phenomena could be treated in strict mathematical terms. The universe was a machine. From this mechanical order a new foundation of physics—field physics—later evolved.

On the philosophical side, the concepts were definitely changing. It was recognized that there was a unity between nature and man or between object and subject. A typical example of this unitary thinking in the eighteenth century was the German philosopher, scientist, and poet Goethe (1749-1832), a Heraclitus modified by 2,000 years of social evolution. His thoughts were of unity process form and expressed a fusion of objective knowledge obtained by the observation of nature and subjective knowledge gained by personal experience.[4]

In the nineteenth century, two great philosophers, Marx (1818-1883) and Freud (1856-1939), attempted unitary thought much before the unitary principle was discovered. Although their thought was one of process, both were forced, by different circumstances, to use static concepts to reach their goals. Marx distorted the tension which is inherent in all processes into the absolute dualism of good and evil to fit the role of proletariat in the class war against individualistic capitalism. Lacking the comprehensiveness essential to any universal theory, Marx doctrine failed. Dogmatic, static concepts could not facilitate the process of history.

Marx morality was concerned with social relations rather than the control of instinctive life. Freud's morality, however, challenged the supposed supremacy of consciousness without knowing how the formative process worked in the brain. He used concepts, such as those of "Super-Ego, Ego and Id," representing broadly consciousness, perception, and instinct, that were not only dualistic but static in the sense of lacking any coordinating principle of development. Although inadequate by scientific standards, his theories were remarkable for the time, particularly the discovery, and sometimes cure, of "neurosis", a condition arising from conflict between instinctive and deliberate thought in dissociated humans.

Only in the late nineteenth century, with Clerk Maxwell (1873) and Pierre Currie (1894), was the conception of a natural formative process definitely established as a one-way process of transformations leading toward the elimination of differences in a stable end-state. By the early twentieth century, unitary thinking was in the air and springing everywhere. The unitary field and the unitary principle were already implicit in the work of Kinji Imanishi (1941),[12] a Japanese scientist contemporary of Whyte. He recognized that our world is a world of things which we perceive as forms, and considered nature as a dynamic relationship of equivalent forces, a balance of mutual influences.

But the glory of the scientific definition of the unitary principle and its application to physics and biology belongs to Lancelot Whyte (1949),[3] who dedicated his entire life to the development of the unitary theory. He is the father of unitary science. A conceptual extension of Whyte's unitary field was later introduced by Baranski (1956),[5] which, when applied to quantum memory organizations, permitted the unitary description of genetic and psychological processes.

Looking back at the development of the formative process *per excellence*, the human thought, Heraclitus sounded the alarm early on with his proposal of an ordering process underlying physical nature and human nature. But it was already too late to save the human thinking process from disaster. Ideas of permanence, such as preservation of life, economic survival, and reproduction, had already dominated the human intellect for centuries before Heraclitus came to the scene. And the temporal relation he proposed was just too difficult for the developing brain to grasp. As a result of this situation, a paradox exists between the history of life, where development is primary and permanence secondary, and the history of thought. Whyte called it "the intellectual tragedy of man."

But the intellectual tragedy continues today despite the availability of unitary science for over half a century. Whyte and Baranski, convinced of the power and veracity of their theory, were very naïve in assuming that unitary concepts would be accepted by all humanity by the turn of the twentieth century. They highly underestimated the power of the resistance forces. Instead, the theory was neglected, not primarily by biology but by physics itself.

How can the unitary principle, the most general and simple quantum field theory ever conceived, in effect a possible universal law applicable to biology and medicine, and to all the sciences, be disregarded by physics in favor of highly complex, mathematically-derived, alien theories with no anticipated applicability to human beings? Fortunately, not all physicists think that way. Among the dissidents stands out Michael Conrad, who in attempting to unify general relativity with quantum mechanics developed a quantum field, the "fluctuon model," similar in many ways to the unitary field.[13] He strongly believed, as minorities of physicists do, that the processes of organisms are extreme expressions of the true underlying physics of the universe.[14]

Although current physics appears to have accepted the irreversibility concept, the classical field is still one that does not allow the development of living structure. Without the possibility of brain development, it is not surprising that there is no room for free will, no explanation for human constructive activities, and a need for *vitalism* in the conception of contemporary physics.[15,16] Only a field with two persistent tendencies, such as that proposed by unitary theory, cooperating in an appropriate environment, can lead to living structure.

Although the description of mental processes given above is very incomplete and anatomic knowledge of quantum memory structures and of all postulated memory areas is still not available, unitary theory proves to be powerful enough to describe our thinking process. It makes us realize that, at bottom, psychobiology is physics. No matter how painful it will be for present physics, unitary theory deserves to be tested and accepted if this proves to be the case.

The symbolic systems of language

In order to communicate the experiences and dominate the environment, our ancestors have ascribed a word or name to every existent thing. Later, they also devised names to describe aspects of, or relations between, things. From this deluge of words a few thousand languages have originated. It can thus be said that we live in two worlds: the real world where things exist and events occur, and the "verbal world," the world that comes to us through words.

In each language, these two worlds must stand to each other as a territory stands to a map representing it. When close correlation exists between the world outside and the world inside our brains, we know beforehand what to expect and we can get work done. But when false maps are given to us in childhood, or when we misread true maps, confusion, shock, and even tragedy may arise.[17]

Spoken words have their origins in primitive cries, which were gradually separated out from their contexts and became *symbols* for particular objects or situations. Symbols, however, are independent of the things symbolized. We can ascribe names to non-existent things and manipulate symbols even when the things they stand for cannot be so manipulated. The word symbols in memory were eventually recorded in script through the development of concepts. The form of these concepts guided the related processes of unspoken thought.

Today, most of our knowledge is received verbally or in written form. We receive the daily news from television reports, radio reports, newspapers and conversation, and we also get information from books, magazines, and movies. Overall, very little knowledge comes to us at first hand, directly from observation of nature. Words, therefore, are at the center of our lives.

Words stand for things but are not things in themselves. The only relation of words to things is the structural correspondence of verbal maps and nonverbal territories. Limitations imposed by language characteristics and by use of static concepts make the object of our experience not the *"thing in itself"* but something less. This is particularly true with things still in process. Words cannot express a changing thing in its entirety; even if they could at a given moment, a moment later the thing would have changed so much that our description of it would no longer be accurate.

Words and concepts are abstractions of memory organizations achieved at cognitive level. Their compacted patterns, however, are not completely devoid of motor and perceptual content. For one, the refined cognitive motor processes keep the sequences of hormone secretion under control and this lessens, but does not abolish, non-specific emotion. Some emotional tone is evident in every concept, and every word is a sound capable of arousing and communicating emotion.

The meaning of words is derived from their context, which is the entire matrix of social life within which each word gradually emerged as a symbol of a particular situation. Meanings therefore are not in the words but in us, in the interpretations and judgments we make when comparing present experiences with past cognitive processes. Because of the complexity involved, it is unlikely for a word to have the same meaning twice. Occasionally, we may interpret meanings of remarks that people have never intended. In these situations, we are confusing abstractions which are in our memories with specific events in the outside world and taking the abstractions as real.

Our system of language was sculpted mainly to communicate our everyday needs. The system, however, is too restricted in its degrees of freedom. It must proceed in one dimension from word to word and cannot describe in one stroke the intricacies of structural transformations and therefore the qualities of our feelings. We must express these qualities in other ways, by manipulation of tones of voice and changes in rhythm or by using connotations, analogies, and other affective devices.

Languages are communicated from a motor system of a person to a sensory system of another. We have been discussing the communication by way of patterns of acoustic energy. This communication system works because our voice apparatus (larynx, pharynx, and mouth) produces sounds in the frequencies of 85-155 Hz for males and 165-255 Hz for females, which are well accommodated within the lower end of the hearing spectrum of 20-20,000 Hz. Persons unable to produce these sounds cannot initiate, and persons unable to receive them cannot capture this type of signal.

Another type of language uses the motor system of hands and face of a person to initiate signs which are received by the visual system of another. In this case, patterns of light energy are involved in signal transmission. A similar but less sophisticated type of body language was used by humans before the spoken language became available. This type of language is therefore very basic. The visual system is also normally involved in the capture of the written language, and this language can be modified for capture by the tactile system in persons with visual impairment.

SELECTED REFERENCES

1. WHYTE LL. Time and the mind-body problem. A changed scientific conception of process. Rhein-Verlag, Zurich (1952).
2. MATSUNO K. Protobiology: physical basis of biology. CRC Press, Inc., Boca Raton, Florida (1989).
3. WHYTE LL. The unitary principle in physics and biology. The Cresset Press, London (1949).
4. WHYTE LL. The next development in man. Mentor Books, New York (1950).
5. BARANSKI LJ. Scientific basis for world civilization. Unitary field theory. The Christopher Publishing House, Boston, USA (1960).
6. NUSSINOV MD, MARON VI. Impulse paradigm of self-organization of matter in the universe, and nanobiological principles. Nanobiology 2, 215-228, 1993.
7. SMETACEK V, MECHSNER F. Proprioception: is the sensory system that supports body posture and movement also the root of our understanding of physical laws? Nature 432, 21, 2004.
8. BROWN JS. Laws of form. The Julian Press, Inc., New York (1972).
9. WHYTE LL. The universe of experience. Harper & Row, Publishers, Inc., New York (1974).
10. ZEKI S. Abstraction and idealism. From Plato to Einstein: how do we acquire knowledge. Nature 404, 547, 2000.
11. ROBINSON TM. Heraclitus. Univ. Toronto Press (1987).
12. A Japanese view of nature: the world of living things by Kinji Imanishi. Pamela J. Asquith, ed., RoutledgeCurzon, London (2002).
13. CONRAD M. The fluctuon model of force, life, and computation. Appl. Math. Comp. 56, 203-259, 1993.
14. CONRAD M. Origin of life and the underlying physics of the universe. BioSystems 42, 177-190, 1997.
15. Physics, philosophy, and scientific progress. Physics Today 58, 46-48, 2005.
16. ELLIS GFR. Physics, complexity and causality. Nature 435, 743, 2005.
17. HAYAKAWA SI. Language in action. Harcourt, Brace and Company, inc. New York (1941).

CHAPTER 8

THE NATURE OF PHYSICAL REALITY

One should never seek to set limits to human understanding.
Lancelot L. Whyte, 1948

We are complex living and thinking beings existing within a much simpler physical universe that we call nature. Nature exhibits patterns such as stars, clouds, snowflakes and rainbows, and therefore it obeys laws. It follows that for us to survive in such ordered universe we must constantly adapt to its laws.

Although our organisms obey blindly to the natural laws, as they are manifested in the terrestrial environment, we know only the laws formulated by the physicists based on their top-down observations and measurements of the external world. The world we know is therefore human-constructed. Present-day physical laws may thus be different from the unknown natural laws.

We know the world around us through our senses, which are very limited in their capacity. They select just a tiny but relevant portion of the total and ever changing reality. The screened reality is then modified by the structural transformations inherent in the sensing, perceiving, and conceptualizing processes. As a result, the world as it appears to us is no more than a truncated part of the momentary reality, and this part does not correspond exactly to its counterpart in the outside world.

Over the last three centuries, the organic limitations imposed by the visual organ on the acquisition of knowledge have been partly offset by the use of ever more sophisticated microscopes and telescopes, which have greatly extended our visual range; and, more recently, the concerns regarding the veracity, or lack of it, of our acquired knowledge have been clarified by unitary field theory.

The unitary postulation of a continuous one-way developmental process from the physical to the biological and to the cognitive realms of reality makes the classical artificial divisions of subject-object and body-mind obsolete. There is only a single structural process in operation in the whole universe, the unitary field process, and our perceptions and conceptions are merely creative manifestations of this process.

But while our perceptions are in close relationship with our environment and so correspond quite well to the outside reality, our conceptions of reasoning and learning, which ordinarily greatly supplement perceptual processes, can be instrumental in disjoining humans from their environment.

Cognitive aspects of reality

The high complexity of living systems prevents them from separating completely from their environments. An organism and surrounding environment must function as one system. In the case of the central nervous system, this means that there must be a close correlation between our perceptual-conceptual organizations and the structural organizations of our environment. Only under these circumstances can our concepts be veridical, our knowledge valid, and our behavior capable of properly developing and directing our environment.

Adaptation of mental processes to empirical situations is a lifelong process that starts in our early years. As we grow older, our concepts become more closely conformed to their empirical referents and, as a result, our perceptions become increasingly more discriminative and our behavior increasingly more veridical.

Our thoughts and behavior are under the control of the central thinking process, which is free to order and reorder conceptual-symbolic sequences to adjust them to the environment. But this organizational process can also generate new concepts of its own, bearing no relation to the environment. Thus, the freedom of memory assembly makes the human thinking process a two-edged sword: a powerful means of adjustment or an equally powerful means of maladaptation.[1]

Individuals therefore have the capacity of developing concepts, such as beliefs, attitudes and values, whose empirical referents are nonexistent or partially existent. Saying it another way, humans or cultures can supply to the environment attributes which do not empirically exist. They exist only in their imagination.

In the description of the human social history given in the previous chapter, we have encountered two such unveracious concepts—the concept of *god* in religion and the concept of *quantity* in objective science. The implications surrounding the first of these concepts, an eternal god guiding us in life and rewarding us with immortality, carry with them very powerful emotional components. This appealing but irrational conception has become deeply rooted in the tradition and it is now a potent factor of cultural disorder.

In all personal experiences and in most observations of the external world there is a continual transformation from earlier to later states. Yet the concept of an eternal god does not involve any element of fundamental transformation. It is a subjective, static concept in a world clearly in endless process.[2]

There is only one universal principle at work: a tendency from structural asymmetry to symmetry. It is this formative process that has created the stars, the earth, and all the life existent on it, by a continuous process of structural transformations. These transformations constitute the rings of a long and multi-branched invisible chain connecting everything with everything else. Contrary to the views of monotheism, unitary science therefore asserts that our mental processes (consciousness) are not separated from our environment, our ideas do not exist in a world of their own, and our personalities (souls) are not distinguished from nature or from our transient physical frame.[2]

We belong to the universe around us and can be neither separated from it nor understood without it. Our physical and mental life is not a separate product brought about by special agencies; it is just a particular example of the operation of the universal formative process.

The other most important unveracious concept, this one in the field of science, is that of quantity. This concept formed the basis of the first heuristic method, the quantitative method. Numbers, however, are just symbolic structures (quantitative concepts) and therefore do not exist in nature. Man puts quantity into nature by applying the concept of measurement.

Measurements and the conceptions derived from them (force, momentum, strain, cohesion, mathematical laws, etc.) are only abstractions and we tend to invest these synthesized abstractions with causal properties that reside only in the empirical referents themselves. So, we tend to think of a force as something empowered with dynamical properties when, in fact, it is merely a mathematical equation derived from mass and kinetic measurements. Measurements are functions of structure and not the other way round. We should therefore be careful not to attribute causal properties to abstract symbols and make them more representatives of reality than reality itself. [1]

The quantitative method is not the essence of the scientific method. It is just the means of achieving more veridical conceptual-environmental relationships by performing ever more accurate measurements. Quantity, however, contains no general principle of form, order or organization and therefore cannot describe biological or mental structure. The method has been quite efficient as an instrument of research, but after four centuries of increasingly intense application in all fields, its usefulness is becoming exhausted. By itself, quantitation can no longer guarantee progressive development of thought.

We must recognize that the basic structure of nature is mostly relational and not quantitative. We must develop a complementary non-quantitative process of measurement capable of describing the structural patterns themselves, that is, the whole molecular arrangement of structure and not only percentages of their components. In this respect, form and order must be more important than quantity and are certainly closer to what is observed. This challenge belongs to the new heuristic method, the unitary method.

A great amount of work is now facing the human race and we are still far from being ready to approach it. This delay is of major concern, particularly in the field of medicine, where the new investigative method has the best hopes of success in the conquest of disease.

Perceptual aspects of reality

Early in the course of animal evolution, sunlight became a strong source of normalizing distortion threatening the continuing development of the life process. Since the light stimulus could not be avoided, the normalizing process was compelled to develop specific structures which could facilitate normalization under these circumstances. These structures came to constitute the visual system. And soon thereafter the eyes became the main port of entry of outside information to the brain.

Light comprises a tiny fraction of the total electromagnetic radiation produced by the sun, but it is a fraction that reaches earth in significant amounts. Like all solar radiation, light is a form of energy that is capable of "traveling" through space without significant loss. At present state of physical knowledge, nobody is quite sure how the traveling process is accomplished and how light "interacts" with matter. Unitary theory, which has proved to be so powerful in the description of biological and mental structure, should be capable of shedding light on this subject, but before it can be of use it must be applied to physical asymmetries (polar, axial, distributional) which are known to be involved in energy phenomena.

The most advanced concepts available on light-matter interactions are those of quantum mechanics. They tell us that when light is moving through space it behaves like waves but on encountering matter it behaves like particles, which are called photons. Classical optics, the study of light-matter interactions, is based on these concepts. Unitary theory, however, does not share them. The particle-wave model neglects the extended character of the pattern and the formative tendency (the development of symmetry) which must exist in any interaction.[3]

Particles and waves represent invariant spatial forms that are conceived by the brain as possessing objective reality. To support their existence, a theory must be based on a space-time coordinate system. Unitary theory is defined by conceptions of asymmetry and succession and therefore waves and particles disappear from consideration. In the unitary view, light rays are just patterns of structural asymmetry made up of aggregates of quantum field particles. There is no distinction between waves and particles because they are the same thing. Unitary theory is thus wider in scope and contains within itself all wave and no-wave theories.

However, for the reasons given above, unitary concepts are not developed fully to allow a unitary description of the different light phenomena observed in the outside world. Quantum mechanical concepts, which provide a very accurate description of unitary structure, within defined limits of validity, will be followed instead.

In quantum terms, light waves are composed of coupled electrical and magnetic fields oscillating in synchrony at very high frequencies. They interact with atomic structure, particularly with its lighter components, the electrons. When light of ordinary intensity encounters an electron, the electrical field of the wave exerts a force on the electron, making it oscillate at the same frequency, but not necessarily in phase with the light wave. The speed of the electron oscillations is very low and therefore the influence of the magnetic field is insignificant. The amplitudes and phases of the electron oscillations, brought about by propagation of the light waves, confer to a material its optical properties.[4]

On interacting with atomic structure, the flow of light can be altered in several ways: it can be *absorbed* (captured) by materials, *reflected* (bounced back) into the same medium by surfaces, *refracted* (bent) at the boundary between media of different densities, *diffracted* (bent and spread out) as it passes through slits or around the edges of an object, *scattered* (deflected in all directions) by non-uniformities in a medium and *dispersed* (spatially separated) by refraction into its component wavelengths. Refraction, diffraction, scattering, and dispersion are all related phenomena.

The electromagnetic energy of light comprises waves ranging from 700 to 400 nanometers in length, the wavelength being the distance between two successive crests or troughs. The energy content of waves varies inversely with their length. Small waves oscillate faster than larger ones because they contain more energy. In general, on traversing transparent materials, high-energy photons propagate faster and interact more strongly with the electronic structure than their lower energy counterparts.

When light is allowed to pass through a triangular glass prism, one of the first experiments performed by Newton over three hundred years ago, differences in velocity and behavior among the light components ultimately leads to their separation (dispersion) into the so-called light spectrum. But while the whole light ray that enters the prism is white, its out-coming separate components are multicolored, ranging from red to violet. Although the light spectrum is a continuum, it has been artificially divided into seven colors, those of the rainbow.

We may now ask where, in this mental experiment, color came from. It must have been added inside the prism, but this does not appear likely since the glass material is itself transparent. If this were not the case, why don't we see the dispersed rays also white? The answer to these questions is that color is not in the light rays. There are no white rays, red rays or any other color of rays. *Color* is not a property of light or of objects that disperse or reflect light. It is a biological phenomenon, a sensation manifested in the perceptual centers of our brain, from information transmitted by our eyes. We see light through color. The external world is colorless. Reality, thus, is not what it appears to be!

But something else is contained in the light spectrum: an intrinsic distributional order. The red-looking rays, the least energetic ones, are always found on one extreme and the violet-looking rays, the more energetic ones, always occupy the other extreme of the spectrum. The myriad rays in between are seen by us to be distributed in the following order, starting from red to violet: orange, yellow, green, blue, and indigo. The green-looking light rays invariably occupy the middle of the spectrum.

In the vision process, light photons must interact with the electronic structure of the retinal pigment rhodopsin. The absorbed energy, converted into electron motion, leads to a series of conformational changes in the rhodopsin molecule that ultimately trigger a nerve impulse to the brain. These changes take place within strict energy limits: too little energy is inadequate and too much energy breaks chemical bonds and forms ions and tissue-damaging radicals. The functional energy span of the rhodopsin complex, approximately 37-75 kcal/mol, is closely met by a wavelength range of 700-400 nanometer.[5] Thus, the light spectrum to which our eyes are sensitive was dictated by the energy requirements of the human visual pigment.

Light enters the eye through a millimeter-sized round opening, the pupil, and on crossing the lens and other structures is coherently refracted/dispersed before striking the retina, the multi-layered neural surface lining the back of the eye. The retina has the shape of a bowl, or of an inverted dome, and its surface is only about 25 cm^2. In a certain sense, the retinal surface may be regarded as the roof of the body, but it is a roof not concerned with solar energy; it has evolved to receive information carried by the light rays after their interaction with the objects of the external world.

To fulfill such specialized function, the retinal microroof must possess a very peculiar structure. Indeed, it displays an enormous array of quantal "light panels" abutting a hundred million tall "chimney-receptors" of two different sizes (large cones and smaller rods), asymmetrically distributed throughout the roofscape. It is a construction marvel unrivaled by any structure that man has ever built.

The ornate retinal architecture has been known for some time, but its functional significance has evaded our understanding. But a geometrical model recently proposed by Gerald Huth, now under scientific scrutiny, appears to unveil some of its secrets.[6] The model is capable of explaining, in different terms, all previous theories but rejects the classical notions that the eye functions as a photographic camera and color is detected by three classes of cones. The retina, it asserts, is located not on the image plan but on the focal plan of the eye, and there is only one class of cones. The determinant of color is considered to be spatial in nature.

Specifically, light wavelengths are defined by the absolute and relative sizes (diameters) of cones and rods and by their spatial order. An ingenious distribution of two receptors differing in size and number can geometrically define with precision three, and only three, wavelengths or narrow bands: two juxtaposed cones define 700 nm (red) and two juxtaposed rods define 400 nm (blue), the extremes of the light spectrum; and one cone juxtaposed to one rod precisely defines, *in actual dimensions*, 550 nm, the middle of the spectrum (green). The all-cone region is located in the center of the retina (fovea) and the all-rod region is located close to the periphery. There is thus a geometrically engineered spatial order for photon reception.

The cones and rods stop the waves in their flight and directly transfer the photons downwards to acceptor sites located between their bases. So the retina primarily registers not the electrical fields themselves but their intensities, the rates at which radiant energy is transferred per unit area. The three retinal bands above described are precisely located at peaks of high density of photon receptors. The photons are then carried to the retinal pigment layer where they are absorbed.

To the light rays entering the pupil, the retinal surface appears like a concave target, with the bull's-eye in the center (the red-sensitive fovea) and two concentric circles or very narrow circular zones, one in the periphery (blue sensitive) and another somewhere in between (green sensitive). The latter defines and fixes the mid-visual band. In color terms, these three circles correspond to red, blue and green—the *primary colors*.

The retina is thought to detect the intensity or brightness of light and its phase at these three target circles and to compare light intensities and phases on both sides of the green circle. The whole light spectrum is defined this way. An outline sketch of a "grey-scale" image of the object is first formed in the cellular layers of the retina from foveal information and a complete image is then assembled. This synthesized information is then carried in the fibers of the optic nerve to the brain, where color and the object are perceived.

The geometrical model of Huth reveals the beauty of the developmental and adaptive processes of biological structure which, over millions of years, led from simple photoreceptors to the chambered eyes of marine animals and finally to the lensed eyes of terrestrial animals.[7] It also reveals the inadequacy of wave and information theories to fully explain the vision process and the generation of color. A more powerful conception of the photon and a theory which takes historical events into account are required.

In unitary theory, the photon is not an "elementary" particle. It is a varying pattern of electromagnetic energy which, at a biological receptor, is transformed by the normalizing field of the organism into receptor-specific configurational quantum field patterns, varying from moment to moment. Each quantum field pattern has its own level of structural asymmetry which introduces a momentarily varying degree of normalizing distortion.[1] Using wave-theory terminology to describe a unitary quantum field structural pattern, it could be said that frequency is related to the varying structural configuration and intensity to its degree of normalizing distortion.

In unitary terms, the function of the photochemical visual substance is to stop each photon, slowdown its speed and transform it (disperse it) into its varying quantum field structural components. Through the polarization-depolarization processes of molecular structure associated with the nerve impulses, these quantum field particles are transported from the nuclear particles of one molecular structure to another, first in the optic nerve and then in other sensory pathways. They finally reach the reticular tropistic and perceptual structuring centers, decoupling and carrying with them similar quantum structures previously recorded in the memory areas they crossed, at each level. Only in the structuring centers, during the temporally varying formative processes of quantum structural aggregation, where structural transactions correspondent to present and past experiences are effected, can color be sensed and objects perceived, as described previously.

If unitary theory is correct, the retina does not supply the whole color information to the brain or the outline sketch or full image of an object. The color of an object is not defined solely by the light it reflects. Part of the information that goes into the color sensation is provided by constantly updated "color" records contained in our perceptual memory organizations. And, as far as it is known, the neural retinal structures do not possess the required physical conditions to support mental formative processes which would be required for image assembly. Unitary conceptions, therefore, move most of the assumed retinal functions to the brain.

By whatever processes, combining the three primary colors—red, green and blue—in the proper intensities gives us the full spectrum of color sensation. This phenomenon, or some variation of it, is explored by the electronic industry in our TV and computer screens and printers. This chapter, however, is concerned with our perception of the real world as revealed to us by light and it is this aspect of color we must address.

When, and only when, all the components of sunlight enter our eyes we see white color; if "light" is emitted from an artificial source, whose spectral composition is necessarily different from that emitted by the sun, we see that light not white but somewhat colored; if isolated sunlight spectral components are received by the eyes, we see the color correspondent to their narrow wavelengths; and when we receive a mixture of wavelength components we see the hue correspondent to their combination.

Some materials have the ability to reabsorb some or all of the spectral components of light. The components not absorbed are either transmitted (by transparent materials) or reflected (by opaque materials). So, an object can be green because it *emits* green photons (a green traffic light) or because it *absorbs* the red and blue light components of the spectrum and reflects only green photons (green car). If all components of visible light were absorbed by a material, it would look black to us. After this simple background, we may now initiate the description of the "seen world" surrounding us.

Because we live within the solar system, on the surface of a planet surrounded by a 100 km thick atmosphere, some of the sunrays on their way from outer space to the earth interact with the randomly distributed air molecules, mostly nitrogen and oxygen (diameter ~ 0.3 nm), and with water droplets and other airborne particles if they are present.

When a particle smaller than the wavelength of light, as nitrogen and oxygen molecules are, is hit by a light wave, it creates a new wave of the same frequency but different phase which propagates equally in all directions. This scattering process is wavelength-dependent and occurs not only in gases but in liquids and glasses as well. In pure air, violet wavelengths (few in sunlight) are scattered three to four times more effectively than red wavelengths.

As a result of this selective scattering of high frequencies, the lower atmosphere is illuminated by scattered blue-indigo-violet light. From all these frequencies, our eyes are most sensitive to the blue. Thus, for an observer on the surface of the earth looking above, the sky looks blue. If the observer were standing above the earth atmosphere, the sky above his head would look black. It is the size of the atmospheric molecules that makes the sky look blue. And it is more intensely blue when the air is dry and dust-free. The red, orange and yellow wavelengths of light are minimally scattered and are able to pass through the atmosphere and reach our eyes. Since the sun is most rich in yellow light frequencies, their dominance is responsible for the yellowish color of the sun.

But the sun is not always yellow or the sky always blue. Their colors change with the time of the day. During the course of a day the earth rotates once on its axis. All celestial bodies, stars and planets included, appear at the horizon to the east of a particular place. They then cross the sky and disappear at the horizon to the west. The most significant and noticeable of these events is the rising and setting of the sun.

At sunrise and sunset, sunlight hits the spherical earth at a narrow angle from the horizon line and therefore takes a much longer path through the atmosphere than during the middle part of the day. Because an increased amount of violet, indigo and blue light is scattered out of the sunrays along the way, the light that reaches an observer early or late in the day is reddish-yellow.

The water droplets that make up the clouds are much larger (diameter $\sim$ 10-20 μm) than the air molecules that scatter blue light. The clouds scatter and reflect all the visible colors of light that strike them, so they appear white, at least from the top. But if the cloud is thick enough, the light rays may not penetrate completely through the cloud, so it appears grey, or even dark, from the bottom.

But from all the natural color phenomena, it was the legendary rainbow that most strongly impressed humans, in every culture on earth. It appears as a colored arc across the sky when falling water droplets are illuminated by sunlight from behind the observer. In Greek mythology, it was considered to be the path followed by the goddess Iris in her occasional trips from heaven to earth to communicate instructions from Zeus to men.[8] The term "iris" is now used to define the membrane responsible for the color of our eyes.

Water and sunlight are both needed to form a rainbow, but to understand its formation we must consider the interaction of light with individual raindrops and not with the whole cloud. The light is refracted as it enters the anterior surface of a spherical raindrop, reflected off its posterior surface, and again refracted as it leaves the anterior surface of the drop to reach the eyes of the observer. Violet light is refracted to an angle greater than red light but the intermediate reflection turns the spectrum over and the red light ends up higher in the sky forming the outer color of the rainbow. In some water droplets, two reflections occasionally occur forming an outer or secondary bow with the colors in reverse, which is separated from the primary bow by a dark band. Obviously, a rainbow is an optical and mental phenomenon caused by the dispersion of light into its colored components by a given number of raindrops and therefore does not actually exist at a location in the sky.

The color of an object can have different causes. We have given examples of scattering, reflection, and refraction and we should now consider light absorption. In this regard, the structure that absorbs most of the sunlight received on earth is water. When seen inside a glass, water appears to be transparent; that is, it does not appear to absorb any part of the visible spectrum. But when seen in large masses, as in deep lakes, rivers and oceans, it has a pale blue appearance.

The blue color of water results from selective absorption of red light by a unique mechanism: the red photons are used to excite molecular vibrations in the water structure. The water color, therefore, is related to the intrinsic motions of the water molecules, which are powered by a permanent electric dipole moment, as described in chapter 3 and elsewhere.[9]

In vapor water, the nuclear motions involve mostly stretching and bending of the O-H bonds. Their vibrations absorb mostly in the invisible infrared, making atmospheric water vapor by far the most important greenhouse gas. But there are also weak bands of absorption in the middle part of the visible spectrum associated with four to eight quantum overtone transitions of the O-H stretch,[10] which behaves like a taut string of a violin.

In liquid water, the stretching of O-H bonds is somewhat diminished by hydrogen-bonding between molecules, and some degree of order also imposes overall motional restrictions. As a result of the lowering of vibrational energies, liquid water is less colored than vaporized water. The vibrational energy of the stretching fundamentals is just high enough for a four quantum overtone transition to occur with absorption at the red edge of the visible spectrum.[11,12] Still lower vibrational energies are found in heavy or deuterized water, whose hydrogen atoms contain one extra neutron in their nuclei. But the more massive hydrogen atoms negate the occurrence of quantum overtone transitions in the visible region, and this makes deuterized water colorless.[12]

Just a few narrow bands of low intensity red absorption are enough to convert white light into pale blue light in the water interior. To reach the surface, where it can be observed, the blue light is then scattered by particles suspended in water. The blue color of deep lakes and seas is more intense than that of shallow rivers. Blue skylight is also reflected at the water surface making it look bluer.

But the predominant origin of color in the natural world is the selective absorption of visible light. The object absorbs certain wavelengths of white light, and we see what is left over. This abstraction mechanism is responsible for the color of gem stones, plants, fruits and vegetables, grass, flowers, corals, animal skin, and blood.

According to quantum mechanical concepts, light absorption by organic molecules involves transitions between ground and excited electronic states. But, for a substance to be colored, the energy difference between the orbitals involved in the electronic transitions must correspond to that of a visible photon. These conditions are realized when the backbone of the organic molecule contains somewhere between eight and fifteen sequences of carbon atoms linked to each other by alternating single and double bonds.

Double bonds next to each other can conjugate (joint together) and delocalize half of their electrons over all the atoms. These delocalized electrons, held in molecular orbitals, can absorb visible light and be the source of color. Conditions for light absorption may also be obtained by way of electron-donor or electron-acceptor groups that pump electrons into or out of the conjugated system, respectively. In certain organic molecules, light absorption is restricted to certain functional groups, called *chromophores*.

Because energy is quantized, it could be assumed that the absorption peaks in a spectrum would be very sharp. But, as we have seen in the case of water, the atoms in a molecule can rotate and vibrate with respect to each other. These vibrations and rotations also have discrete energy levels, packed on top of each electronic level. As a result of the superpositions of rotational and vibrational transitions, a combination of overlapping lines is obtained which appears as a continuous absorption band.

Chemical substances that absorb only certain wavelengths of light are called *pigments* and *dyes*. Pigments were first synthesized by bacteria and plants for the purpose of absorbing light energy for their own metabolism. Later in evolution, plants and flowers produced colored compounds, not for absorption of light but for selective reflection of specific colors. The colors were, and are, used to attract insects and birds for pollination purposes. In animals, devoid of photosynthetic capabilities, colored pigments are primarily used to block the propagation of solar radiation into tissues. This is certainly the case of black *melanin*, which absorbs the potentially damaging ultraviolet sunrays that penetrate our skin and may also be the case of red *hemoglobin*. It is very unlikely that chance has dictated the color of our blood.

In thermoregulated organisms, the heat produced in metabolism must be dissipated into the usually colder environment and the heat generated in the environment must be prevented from dissipating into the organism. Most of the heat energy is contained in the invisible near-infrared and visible red parts of the solar spectrum. About half of this energy is absorbed in the epidermal layer, the remaining reaching the dermis, where the skin microcirculation flows and the sweat glands are located. The red hem of blood reflects the above energies back into the environment, thus negating the absorption of heat.[13] This blood characteristic reaches special importance in hot environments when dermal vasodilation must be sustained to allow dissipation of metabolic heat by the sweat glands. Under these circumstances, blue blood would be inconsistent with life.

The absorption spectra and respective colors of some organic pigments are schematically displayed in Fig. 8 (pg. 58). It can readily be appreciated that a green colorant, localized at the middle of the spectrum, must have two absorption bands, one at each end of the spectrum.

The other chemicals with specific light absorption are the dyes, compounds that impart color to other materials. For millennia dyes have been extracted from minerals, several parts of specific plants, from roots to leaves, and even from marine animals and used to color leather, textiles, skin, and hair and in the drawing of art works in caverns in pre-historical times. The most famous of all dyes was probably *indigo,* extracted from tropical *Indigofera* species, woad and other plants. It owes its bluish-red color to absorption in the yellow-green part of the spectrum. Asian countries, such as India, China, and Japan, have been using it for centuries. Today, natural indigo is still used in the production of *denim* cloth (from *"de Nimes"*, a town in southern France) for blue jeans. Pigments and dyes of all colors are now synthetically produced. Pigments form the basis of all paints and dyes are used to color almost everything, from glass to textiles to paper.

Knowing that the red color of a rose is not in the rose does not diminish its beauty. It just changes the perspective. Going one step further, the beauty that a rose manifests should not be regarded as being in the rose, either. The true beauty resides not in the forms themselves but in the formative process of nature that generates them. And the more we know about the workings of nature, the more appreciative of all creation we become.

In the vegetable realm, metal ions are commonly found in pigments, such as magnesium in chlorophyll. Sometimes a combination of metals ions, such as iron, magnesium and calcium, is responsible for changes in the coloration of a given pigment, and other times more than one pigment contributes to color. All these different combinations are responsible for the huge variety of colors found in nature.

In the less complex inorganic world, metal ions present as impurities in otherwise transparent crystals, in concentration of 1 percent or less, are themselves the agents of color. The metal impurities allow electron transitions or charge transfers to be effected by visible photons. This occurs in the great majority of *gemstones.*

When in its pure state, diamond is transparent. Its carbon atom matrix, crystallized under pressure, is so tightly bound together by chemical bonds that no photon in the visible energy range has enough energy to break it. So, diamonds are transparent because they are incapable of absorbing visible light. The same is true for corundum, the colorless matrix of ruby and sapphire, and for the pure crystal of beryl, which forms the basis of emeralds and aquamarines.

The metal ions most commonly embedded in gemstones are those of chromium, titanium and iron, single or in combination. Their valence electrons can transiently be moved by the energy contained in visible photons, and it is these motions that, in a direct or indirect way, are associated with color.

In some gemstones, these electron motions occur between the atomic energy levels of a single metal impurity. In these cases, the coloration depends, among other things, on the physicochemical characteristics of both the "host" matrix and the "guest" impurity, and on the strength of the interaction between them. As a result, the same impurity is capable of generating different colors in different crystals.[11]

This is the case of ruby and emerald, both containing chromium ions as impurity. The corundum matrix of ruby, made of repeat units of aluminum oxide, lowers the two transitional energy levels of its guest impurity just enough to allow the absorption of green and violet photons. This spectral absorption gives ruby its characteristic red with a tinge of blue; the weaker interaction of the beryllium aminosilicate of emerald with the same impurity lowers further the energy levels of the chromium atoms, thus promoting the absorption of red-yellow and blue photons. This pattern of absorption gives emerald its green color.

The intense blue color of a sapphire derives from a different mechanism. It results from cooperative interaction between vestigial amounts (0.01 percent) of two metal impurities, titanium and iron atoms, within a corundum crystal. The energy conditions are such that red, yellow and green photons can pull a valence electron from iron and transfer it momentarily to titanium. Blue photons cannot do that and so sapphires appear blue.

Similar color schemes involving visible photon absorption to move electrons, within or between atoms of resident impurities, are responsible for the coloration of other gemstones. Opal, however, is an exception. It owes its flashes of color to a combination of light refraction and diffraction. In a way, it behaves as possessing a myriad of tiny prisms in its structure, although no prisms are there to be found.

The stone consists of a regular tri-dimensional array of tiny uniform spheres of transparent, hard silica, containing a small amount of water, in a jelly silica matrix containing a different amount of water. The refractive properties of these two media are slightly different and as a result light is bent as it passes from one medium to the other. In addition, light is diffracted as it passes between the spheres. Depending on the inter-sphere spacing, somewhere between 0.2 - 0.3 nm, varying colors of the spectrum are diffracted. When the gemstone is rotated and tilted, light hits the spheres from different angles and brings about a change of colors, on a white or black background. This rainbow effect is known as "play of colors".

The exhibition of rainbow-like colors on a surface is called *iridescence* (from Iris, the goddess of the rainbow). It results from interference or interaction between light waves. In constructive interference, the waves meet in phase and the resultant wave has increased amplitude (and a brighter color); in destructive interference, the waves meet out of phase and cancel each other out in some degree.

When light strikes a two-dimensional assembly of narrow slits, called a diffraction grating, it diffracts and produces a large number of beams which can interfere in such a way as to produce spectra. A similar process occurs in the inter-sphere spaces of the opal structure. It is wave interference that leads to the rainbow effect. Opal is a natural diffraction grating operating in three-dimensions!

There is something magic about iridescence which holds our attention and pleases our senses. This photonic show is far more developed in the animal world, particularly in terrestrial animals. It is seen in the wings of tropical butterflies and in the plumes of hummingbirds[14] but it reaches full splendor in the tail feathers of peacocks. Next to the ethereal rainbow, the stunning beauty of a fanned peacock tail ranks second among the wonders of creation.

In his work *Opticks,* published in 1704, Newton had already related iridescence to optical interference but only in recent years has the photonic structure of peacocks received some attention.[15] Its color mechanisms, however, are still not fully understood. In a peacock tail feather, the most striking color pattern is found in the oval-shaped eye. From the periphery to the center of the eye, the colors reflected are yellow, brown, green, and dark blue. The structural architecture responsible for these colors lies in the barbules, the twig processes along either side of a barb. It consists of melanin rod arrays in a keratin matrix, interspersed by air pockets, forming a three-dimensional multi-layer grating. Among the colored barbules, the number of layers varies and the spacings between the melanin rods are slightly different to allow different wavelengths of light to be filtered and reflected.[8,15]

For colors to be iridescent, they must be of single or very narrow wavelength and must change when the iridescence surface is viewed from different angles. This is achieved by a combination of absorption, reflection, diffraction, and interference. Light diffracts when it passes though the inter-melanin slits and this effect is multiplied by many plans of parallel slits. The different colors correspond to different length scales of the periodic structure. The melanin pigment intensifies the colors by absorbing the nonreflected light.

Up until now we have been describing colored phenomena brought about by interaction of sunlight with inorganic or organic structure. But light is a form of energy and as such it can result from transformation of some higher energy form, such as chemical energy. The light produced by a chemical reaction is called *chemiluminescence*. Valence electrons of specific chemical compounds are first excited to a higher energy level and then, on falling to ground level, light is released without significant heat.

An example of a chemiluminescent reaction is the one taking place in the plastic glow-sticks sold to children in amusement parks. Two chemicals interact to release energy, which is accepted by a fluorescent dye and converted into light. When similar reactions occur within living organisms, the phenomenon is called *bioluminescence*.

In the biochemical reaction, the source of energy is ATP. It reacts with the substrate pigment luciferin, with the aid of the enzyme luciferase, to yield an intermediate complex. This complex combines with oxygen to produce a highly fluorescent compound which gives off light. The forms of luciferin and luciferase differ chemically in different organisms, with the color of light given off varying according to the substrate.

Some 90 percent of deep-sea marine life produces bioluminescence in one form or another and uses it as lure to attract prey or mates and for camouflage. The light emitted is usually blue or green, which are the wavelengths transmitted through sea water most easily. On land, bioluminescence is less widely distributed, but some species of fungi and bacteria, as well as fireflies and glow worms, are capable of emitting light of different colors.

Probably the most famous glow worm is the so-called railroad worm, a larva of the beetle *Phrixothrix hiatus* found mostly in Paraguai and Brazil that produces red light from its head and greenish yellow light from eleven spots on either side of its body, resembling the illuminated windows of a train. The substrate that produces the red light in the railroad worm is apparently the same that lights up the nose of another even more famous creature, the wild reindeer *Rangifer tarandus rubens*, nicknamed Rudolph, an inhabitant of the cold regions of Norway and Russia.[16]

We are connected to the outside world mostly by way of photons and, in a general way, what has been said for light photons is applicable to sound photons as well. However, in the case of light, the sun supplies us with a fairly constant amount of radiation reaching us at a constant velocity, and our eyes adjust to the intensity of light by constricting or dilating the pupils.

The same does not occur with sound. Our ears are devoid of protection against the intensity of sound, and the sources of sound can be extremely varied. Take the case of an explosion. The sudden release of energy is associated with innumerable physical events or sound stimuli occurring in physical time. The biological perception of sound, however, requires a time interval of the order of 0.1 sec for structural aggregation to take place in the centers of the brainstem. All these countless stimuli are therefore contained in a single sound perception. This is the reason why explosions appear to us so enormously violent.[17]

The fourth spatial dimension

The physical world, as we see it and routinely experience it, is spatially three-dimensional. Three spatial dimensions are the maximum that our visual system can see. The visual process involves the eyes-brain (locus of vision) and an illuminated or luminous exterior object, both of which exist within a three-dimensional plane of space (focus of vision). Recently, however, Paul Hollister has called our attention to another spatial dimension, which has been experienced by some of us but hitherto unrecognized as such. He calls it the fourth spatial dimension.[18] This dimensionality is a natural consequence of the unitary field process, and for its proper understanding we must think in process terms.

The asymmetry to symmetry tendency of the unitary field was the *primum movens* for the gradual construction of a hierarchy of structures throughout the universe. This hierarchy proceeded from smaller to larger magnitudes of mass, from simpler to complex forms, from particles to atoms to stars to galaxy formation. This one-way development process originated in the basic energy field (inner space) and expanded outwards (outer space). If we mentally align all these mass magnitudes in the same axis, we can envision a dimension of depth, the fourth spatial dimension.

Living structure also developed from the basic energy field outwards, from particles to atoms, molecules, cells and finally organisms. This biological "world" was assembled on the surface of the planet earth, between the magnitudes of atoms and stars, in the middle of the very small and the very big. Therefore, it can be aligned, at this location, on the axis of the fourth spatial dimension. In the universe as a whole, all magnitudes of mass are structurally layered in three dimensions along this dimensional axis of depth.

While the atom-based physical world is characterized by huge linear size, the molecule-based biological world is characterized by huge structural complexity. Complex organisms, although dependent on their physical surroundings for existence, enjoy enough autonomy to act as observers of the physical reality. When we look at the structure of the universe through our eyes (cone of vision) we are physically outside of the composite whole, standing on the surface of the earth.

With the technological advances of the last four centuries, modern humans can look at the physical structure of the universe through potent telescopes and into the inside of earth and of their own bodies through equally potent microscopes. Only by the use of this instrumentation can the fourth spatial dimension be perceived.

What these instruments actually do is move our eyes from a three-dimensional focal plane of space to another. When we are looking at a faraway galaxy through the Hubble telescope, our eyes are at the level of that galaxy looking outward, and when we are looking at a cell through the electron microscope, our eyes are at the level of that cell looking inward. In both situations, our eyes are moving on the axis of the fourth spatial dimension of mass magnitude, from inside-outside and from outside-inside, respectively. It is during these motions that we are able to see and experience this elusive dimension of depth. As Hollister points out, we have been confusing magnification with magnitude. Magnification is a function of vision, whereas magnitude is a function of experience.

The intrinsic order inherent in the formative process of structure can be mathematically defined with precision. All magnitudes of mass have spin and therefore the spin axis could theoretically be used to precisely align these magnitudes on the axis of the fourth spatial dimension. This should allow the assembly of all three-dimensional layers and parts of the Universe together, allowing us to accurately see the composite structure of the whole universe.[18]

SELECTED REFERENCES

1. BARANSKI LJ. Scientific basis for world civilization. Unitary field theory. The Christopher Publishing House, Boston, USA (1960).
2. WHYTE LL. The next development in man. Mentor Books, The New American Library of World Literature, Inc., New York (1962).
3. WHYTE LL. The unitary principle in physics and biology. The Cresset Press, London (1949).
4. MOUROU GA, UMSTADTER D. Extreme light. Scient. Amer. 286, 81-85, 2002.
5. LUO Y-R. Why is the human visual system sensitive only to light of wavelengths from approximately 760 to 380? An answer from thermochemistry and chemical kinetics. Biophys. Chem. 83, 179-184, 1999.
6. HUTH GC. A new physics-based model for light interaction with the retina of the eye: rethinking the vision process (2005). Complete manuscript on line at *http://www.ghuth.com*.
7. FERNALD RD. Casting a genetic light on the evolution of eyes. Science 313, 1914-1918, 2006.
8. ZOLLINGER H. Color. A multidisciplinary approach. Wiley-VCH, Weinheim (1999).
9. CARVALHO JS. Life: its physics and dynamics. American Literary Press, Inc., Baltimore, MD (2003).
10. BERNATH PF. Water vapor gets excited. Science 297, 943-944, 2002.
11. NASSAU K. The physics and chemistry of color. The fifteen causes of color. John Wiley & Sons, New York (1983).
12. BRAUN CL, SMIRNOV SN. Why is water blue? J. Chem. Edu. 70, 612-614, 1993.
13. SAUTER C. Why human blood must be red. Amer. J. Hematol. 29, 181, 1988.
14. VUKUSIP P, SAMBLES JR. Photonic structures in biology. Nature 424, 825-855, 2003.
15. BLAU SK. Light as a feather: Structural elements give peacock plumes their color. Physics Today 57, 18-20, 2004.
16. KEMP M. Nasal bioluminescence in reindeer, *Rangifer tarandus rubens*. Nature 426, 768, 2003.
17. AXELSSON S. Perspectives on handedness, life and physics. Med. Hypoth. 61, 267-274, 2003.
18. HOLLISTER P. The origin and evolution of the universe, a unified scientific theory (2004). Complete manuscript online at *http://www.origin-of-universe.com*.

CHAPTER 9

THE SOCIO-CULTURAL ENVIRONMENT

*Societies, both animal and human, might almost be regarded
as huge cooperative nervous systems.*
S.I. Hayakawa, 1939

Humans are very gregarious beings. They have a natural tendency to develop interpersonal relationships based on mutual needs and common instincts, emotions and ideas. These relationships are greatly facilitated by their cognitive processes of language. In the course of social development, they have led to the formation of aggregates called social groups. The first social group, the family, was followed by larger and larger social, economical and cultural groups, which ultimately formed communities and nations. As a whole, the individuals and their associations constitute the socio-cultural environment.

The patterns of individual and group relations in the highly complex social world have the same fundamental significance as the patterns of relations of parts to systems and systems to larger systems in the less complex organic world. Both are manifestations of the intrinsic tendency to symmetry of the normalizing process underlying the whole universe. This formative process successively led to the evolutionary development of the metabolic system, which brought about the rise of life and the more complex and much larger social-economical-cultural system from which the human personality and culture have emerged. This last organizational level is the highest that the field process has achieved on this planet.[1]

The socio-cultural environment is made up of human personalities connected by interrelated and interdependent cognitive and instinctual relationships. Each human personality must simultaneously contribute and adapt to his specific social environment. The interpersonal relationships are at the basis of this cooperative process of social development. The formative individual contributes with novel ideas, and the conservative society molds the individual to past experience and accepted behavior and allows him to carry the process further.[2] The socio-cultural environment is thus a school of personalities and can be considered an external cognitive system. The knowledge contained in this system is transmitted from one generation to the other by the tradition.

The various personalities comprising a social group are separated in space but closely interconnected, by personal relations, into a functional whole, the social unit. The asymmetrical character of the personal relations and the cooperative interaction among the individual constituents impart on the social unit the characteristics of a formative process. The social formative process has its particular laws, dependent on the degree of complexity and properties of the structural aggregates, but, as all other formative processes in the universe, it is governed by the same principle of decreasing symmetry, which is manifested in one-way development.

The energy and organization for the origin and development of all social systems was supplied by the normalizing process working as an integral part of the instinctive, perceptual, and cognitive sub-systems of the individual personalities comprising each specific social system. In this view, all systems that sustain our societies, those of agriculture, commerce, technology, communication, culture, etc. are extensions of the individual's intrinsic psychomotor activities of manipulation, locomotion, and language. Countless numbers of organized discoveries and inventions have contributed to the intricate societal formative processes of today. These processes facilitate normalization in the individual and greatly multiply his ability to organize the environment.[1]

All structures built by humans or associations of humans are the result of formative or structuring processes that used individual memory organizations as their immediate directive and developmental agents. In the construction field, for instance, humans first developed tools, shaped stones and wooden boards, invented cement, iron rods, bricks, glass etc., all of which are simple structural aggregates. Then they structured these aggregates into more complex ones in the building of our homes. Then, further structuring of these aggregates led to the construction of towns, sky-scrapers, cities, etc.

Contrary to classical physics, unitary theory is thus capable of explaining not only mental processes but all outcomes or expressions of human creativity. It asserts that a universal structuring tendency, an intrinsic process of ordering, runs throughout all levels, from the physical to the social. It is this formative and creative process that differentiates all structural organizations, from simple to complex.

The human personality, therefore, is not reducible to the properties of the basic field. On the contrary, it has acquired very unique properties by countless differentiations of the field, first from field to cell, then from cell to organism, and then again from organism to personality. It is this specific differentiated organization, where past events are taken into account, which makes for uniqueness or individuality.[1]

The normalizing process is self-facilitating. It always tends to synthesize structures that facilitate normalization. In other words, it always works in the direction of increasing order, even when the environment is non-facilitating. But there are always elements of disorder in individual brains and in social and cultural institutions that may override the normalizing tendency. Under these circumstances, the individual may fall out of step with various aspects of reality with potentially serious consequences.

Sources of socio-cultural disorder

In a given sense, the human personality lives and develops in two worlds, the physical world of stars, mountains, trees, and rain and the social world of other developing human personalities. We perceive these worlds through symbols and communicate with one another by way of symbolic concepts expressed in language.

It was the symbolic process, the understanding by humans that things can stand for other things, which made language possible. Unfortunately for the human race, the intimate connection of physical to mental processes revealed by unitary theory was not clearly understood until the early twentieth century. As a result, the mental symbols were taken as isolated from physical structure and therefore had no direct relation to physical laws.

This isolated condition was born from the dualistic cognitive systems of pre-unitary time, namely the dualism of subject-object, expressed in the philosophical notion of *ontological cut*. The essential function of the symbol was to stand for something—its referent—which, by definition, was on the other side of the cut.[3] Symbol and referent, verbal world and real world, were isolated entities, independent of each other.

Although the symbols are distinct from their empirical referents, there exists in the unitary conception a hidden physical continuity between them. The empirical referents are causally related to the symbols and are thus not separate entities. The symbolic concepts therefore must possess, or progressively attain, a one-to-one correspondence with their respective empirical referents. When these circumstances are met, the minimum of disruptive effects or normalizing distortions are experienced by the individual.

But the characteristics of the human thinking process are such that we are free to construct symbols at will, even if they stand only for abstractions and not for anything in the physical world. This astonishing characteristic of humans is at the basis of their creative activity in the arts, literature, sciences, and in life in general.

But when language is used in the communication process, the symbols that are supposed to represent empirical referents must actually represent them. In the figurative expression of Hayakawa, which ignores process, our abstract concepts ought to stand in relation to things "as a map does to the territory it is supposed to represent."[4]

Sometimes, however, beautiful maps are constructed which bear no relation to the represented territories, the overall structural patterns by which nature is viewed being basically flawed. These conceptual symbolic abstractions are, in reality, devoid of denotable empirical referents, but by attributing causal properties to the symbols, referents are implied to exist. The process whereby a faulty relationship between concept and environment is established is called *reification*.

Reified concepts are not veridical. They do not correspond to the outside reality when investigated by the scientific method or when the concepts are viewed in a larger context. They are, in effect, subtle forms of cognitive disorder. As time goes on, the unveridical reified concepts are structured into attitudes, beliefs and values that selectively affect perception, cognition, and behavior. If the distorted memory structures are not reordered or removed, the individuals, social groups or nations fall out of relationship with their environment, and this disconnection leads to false and distorted learning and behavior and to maladaptation in the long run.[1]

Since most of our knowledge is received verbally, reified concepts are usually acquired from the tradition, social groups or other individuals, but they may also be constructed by the individual, either from direct observations or, more commonly, from his own memory organizations. They result from wrong assumptions, misunderstandings, limited knowledge, casual lies, and many other causes. These unveridical abstractions are thus widespread in society, where they constitute sources of irrationality, confusion, and disorganization.

If recognized, reified concepts can easily be eliminated by the normalizing process. But, characteristically, their users are seldom aware of their presence and disintegrating influence. Over time, the flexibility of thought becomes impaired, and rigid behavior develops. A black-white or two-valued orientation sets in, which blocks or distorts the learning process and further isolates the individual from his environment. This isolation prevents both the selective elimination of distortion-producing conceptual stimuli and contact with conceptual stimuli which could facilitate development in relation to the environment. In this way, the individual or social group is progressively led in a direction away from a valid relationship with the environment.

The static, two-valued orientation is often used by propaganda whereby one group is always right and the other always wrong or one product is good and all the others are bad. This form of overgeneralization serves to block the learning and judgment processes and this is the reason why propaganda is one major source of cultural disorder.

But at bottom, most of the disorder and disunity that characterizes our present civilizations comes from the old dualistic cognitive systems of objective science, philosophy, and religion, which, despite the passage of many centuries and the rise of unitary thinking, are still with us today. So, quantity in science, idealism in philosophy, and god in religion continue to be the most important reified concepts of our time.[1,2] This is so because they influence, in a direct or indirect way, all aspects of our lives.

We have already referred to the quantitative principle as the basic instrument of scientific research. But over the last four hundred years, overemphasis on quantity has led to reification of the notion of mathematical (physical) laws and statistical techniques. They have become things in themselves instead of guides to empirical referents and qualitative concepts. Gradually, the quantitative method evaded the controlling influence of mental processes and became autonomous.

We owe to the quantitative principle great advances in science and technology and considerable improvement in our living standards, but we must recognize that structure, physical and biological, is not wholly quantitative. Basically, it is spatially related and in continuous process. The structural patterns are not describable by quantities alone.

The quantitative method is one of analysis, so it can lead only to differentiation, not to integration and organization of thought.[2] Integration is implicit in formative processes where order is generated. Therefore, the method is intrinsically anarchic. It also neglects form and is thus static. Although technology establishes closer specialized relationships, it does not organize the whole pattern of life in any social group.

In the economical field, the lack of order of the unrestrained application of the quantitative principle has now become apparent. It has converted the Darwinian "struggle for survival" ideas, which underlie the American economy system, into those of "struggle for maximum profits". There are no limits to wealth in the economic arena. Accumulation of money has become the ultimate purpose of life.

In the scientific field, a similar anarchy has occurred. The intense application of the physical laws and statistical techniques has generated enormous amount of information in all branches of science, now awaiting some kind of organization that the quantitative method cannot provide.

What was said for the quantitative scientific principle also applies to the religious and philosophical systems that preceded science. Both contain great wisdom and human values and have made contributions to society of great importance, but they are products of dualistic, and therefore reified, cognitive systems. Both lack the principle of denotability and do not possess a method for testing their concepts against experience equivalent to that of science.[1]

The universal process as source of order

In the unitary view, nature is a system of systems in continuous process. The Milky Way galactic system, the solar system, the earth system, the organism-in-environment systems, the organism systems, and the personality systems are all components of a wider system that we call the universe. A permanent god has no place within a universe in process. And since order can only come from process, no ordering principle or entity can exist outside the universal process.

The scientific understanding of the interrelatedness of things, brought to light by unitary science, gives us a new insight into the workings of the universe and into ourselves. The asymmetrical relations of mental processes are as real as those of physical processes. We can no longer continue thinking that our bodies and brains are made of stuff different from that of physical things and the living process is something given to us by some omnipotent deity. We are no more than hierarchical systems of formative processes in continuity with the universal formative process where all order is contained. Essentially, our thoughts are complex pieces of the basic field emanating from the cognitive processes through the action of the normalizing process.

When veridical concepts connect us to nature, the normalizing process becomes facilitated and our thoughts are suitably ordered. This is the situation of cognitively healthy individuals capable of forming cognitively healthy and health-promoting social organizations. And only under these circumstances can interpersonal relationships lead to mutual understanding and cooperation, the ingredients behind the orderly growth and differentiation of societies.

Life is process and process is development. For the individual, this means development of the human personality, the achievement of a rich cognitive and instinctual life. The supreme aspiration for the individual in society is therefore not freedom but the quality of his or her experience—happiness, beauty, truth.[2] Thus, the concept of development takes value away from the accumulation of money and property. Money can give comfort, but not happiness.

Without proper understanding of the human connection to nature, human culture has been wandering over the ages, from the Stone Age to the Computer Age. The progress has been slow but we have come a long way from pre-religious superstition to science, tribe to nation, food gathering and hunting to industry, and patriarchy to democracy. But we are far from reaching the socio-economical-cultural level that the potentialities of *Homo sapiens* permit.

The fact that humans have been able to civilize themselves shows that some self-regulating development has in the main dominated the history of the species[2]. Disorder is thus not inherent in humans. The normalizing process always tends to work toward order, even when the structural organizations and the socio-cultural patterns of a society are non-facilitating. There are thus reasons to believe that a new kind of social order, based on unitary concepts, may be capable of uniting humanity.

Towards unitary order in society

The society is the basic source of symbolic concepts which individuals acquire, via learning and education, during formation of their personalities. The individuals, in turn, influence the society through new concepts and vistas. This cultural interplay between individual and society makes human thought not arbitrary. Individual ideas, attitudes, values, and actual behavior are all components of a developmental social process.[2]

Under this view, religion, philosophy and science are not independent activities but different expressions of one underlying process, the cultural development of the species. Thus, our thoughts and ideas can only be fully understood when viewed through the process of history.

Since the start, human culture has been built on fear, first of the gods, then of the lord, of sin and death, and fear has never left the mortal humans. Our present civilizations continue to impose still more fear—of job loss, disease, government, terrorism, war. These multiple fears are stored in the mental memories and tend to motivate aggressive behavior. Any future social order must therefore be based on the progressive elimination of fear.

The historical process is a continuous unitary transformation but, in broad terms, we can say that during this process human thought went from the subjectivism of religion to the objectivism of science, from introspective idealism to the objective study of nature. Between these extreme phases, forcing the change, there were the dualistic systems of philosophy. All these cultural phases were supported by unveridical cognitive systems.

None of these disciplines has been able to unite humanity. The religious concepts disregard the trend of history and are too vague and undifferentiated to carry out such a task; the intellectual dualisms of the philosophical and politico-philosophical systems cannot provide the integration of thought required for the establishment of a social order; and modern analytical science, detached from human nature and unrestrained by any type of order, cannot unify.

Instead of bringing order, the rapid advances of science and technology have increased the degree of social disorder. Computers and robotics, for instance, which should have set laborers free of repetitive tasks and increase their leisure time, have taken away their jobs and lowered their salaries instead. And ever-more-sophisticated weapons of highly destructive power are now holding humanity in permanent fear.

The atomic bomb, in particular, is now used as if it were the last cross or cipher in a grandiose tick-tack-toe which no nation wants to win but cannot afford to lose. The strategy of bombing the adversaries faster and more destructively than they can bomb us invites the use of the bomb in a moment of extreme fright, with devastating consequences for all humanity. Certainly, this aggressive stand among nations cannot unite them.

The mechanical processes that led to the development of technology, industry, and warfare can be accelerated by deliberate effort, but mental formative processes, which underlie our intellectual, esthetic, and moral faculties, cannot. The development of new ideas, purposes, and judgments, which are required for adaptation of humans to a rapidly changing socio-cultural environment, is a time-requiring process and cannot be rushed. As a result, the purposely fast development of technological products inevitably led to intensification of social friction.[2]

When a society is disordered, the thinking and behavior of its individual members and social groups also tends to be disordered. Under these circumstances, only appropriate ideas can unite people. And to unite all the people, a universal idea with the force of a religion is required. Such universal idea or principle should be based on scientific truth, unchallenged by criticism and freely accepted by everyone in all countries. The unitary principle contains all these ideological requirements and its proponents clearly saw its relevance as an instrument, not only for the unification of scientific knowledge but for the unification of humanity as well.

At the time when the historical phase of neutral objective science is decaying, it is appropriate, perhaps even inevitable, to replace it with a science that brings subjectivism and objectivism together in a single theory which also takes the whole historical process into account. It replaces the static concept of quantity by the more powerful and lively concept of order, which is capable of overcoming fears, the basic causes of social disorder.

We should start by discarding the primitive and misleading static systems in favor of a *qualitative cognitive system* that is scientific, veridical and adaptable to changes in the social environment. Only under a system of cognitive order can high human values and ethics be achieved and intellectual, moral and social order restored.

Technology has brought the world together, and the world is now waiting to be united. Unitary order based on scientific reasoning, human values and mutual development appears to be the best approach. Only cooperation can possibly lead to economic security, absence of fear, and achievement of high quality human values.

The unification process will be slow and painful. It calls for new educational, political, and economical systems based on field concepts, but their details are beyond the scope of this book, whose main purpose was the scientific explanation of our relation to the physical universe of which we are a part.

SELECTED REFERENCES

1. BARANSKI LJ. Scientific basis for world civilization. Unitary field theory. The Christopher Publishing House, Boston, USA (1960).
2. WHYTE LL. The next development in man. The New American Library of World Literature, Inc., New York (1950).
3. PATTEE HH. The physics of symbols: bridging the epistemic cut. BioSystems 60, 5-21, 2001.
4. HAYAKAWA   Language in action. Harcourt, Brace and Company, New York (1939).
5. WHYTE LL. Everyman looks forward. Henry Holt and Company, New York (1949).

INDEX

absorption band, 100
absorption spectrum, 100-101
abstract concept (*see* psychological
    phenomenon)
achiral (*see* chirality)
action potential, 68
ADP-ATP system, 71
adaptation (*see* process, adaptive)
adenine (*see* purine)
affection (*see* psychological process)
aging, 39, 69
agriculture, 83, 108
algae, 17, 46, 60
aluminum (*see* atom)
amino acid: 20, 35, 37-38, 44-45
  glycine, 38
  l-amino acid, 38-39
  valine, 44
ammonia, 8, 13
anabolism (*see* metabolism)
anaerobic glycolysis, 45
angular momentum, 12
animal, 39, 74
anticipation (*see* psychological process)
antigravitational field (*see* field)
aquamarine, 101
artificial life, 42
artificial organ, 61
astronomer, 30
astronomical time (*see* time)

asymmetric atom, 38, 67
asymmetrical relation: 61-62, 64, 69, 73
  spatial, 25, 27, 28, 31, 40, 64, 68-69
  temporal, 25, 27, 69, 87
  numerical, 27, 74
  social, 83, 86, 107, 111
asymmetry (*see* structural asymmetry)
asymmetry chain, 76-77, 81-82
asymmetry level, 45-46
asymmetry gradient, 41
atmosphere, 11, 13-15, 17, 97-98
atmospheric carbon dioxide, 60
atmospheric oxygen, 60
atom: 3, 6, 8, 11, 14, 23, 29, 36-38, 40, 63, 65,
    67, 69, 85, 99, 104
  aluminum, 14
  calcium, 14, 23
  carbon, 6, 7-8, 37-38, 66, 70, 99, 101
  cesium-133, 30
  chromium, 7
  helium, 4, 6-8, 12
  hydrogen, 3-8, 12, 18, 70-71, 99
  iron, 7, 13-15, 17
  magnesium, 6, 14
  manganese, 7
  nitrogen, 6, 8, 39
  oxygen, 6-7, 13-15, 17-18, 25, 70-71
  phosphorus, 6, 14
  potassium, 14
  sodium, 6, 14

silicon, 6-7, 14
  sulfur, 6
atomic bomb, 112
atomic clock (*see* clock)
atomic matter (*see* matter)
atomic organization (*see* organization)
atomic pattern (*see* pattern)
atomic structure (*see* structure)
atomic time (*see* time)
autocatalytic process (*see* process)
automatic pilot, 81
autonomic nervous system (*see* nervous
     system)
autonomy, 59, 104
ATP system, 43, 45, 102
ATP-nucleoprotein system, 45
ATP synthase, 71
axial asymmetry (*see* structural asymmetry)
axis, 38-40, 69, 98, 104

bacteria, 17, 31, 39, 100, 103
balance, 75
basalt, 14
basic field (*see* field)
behavior (*see* psychological phenomenon)
beryl, 101
big-bang theory, 4-6
biochemical level (see structural level)
biochemical reactor, 35
biochemistry, 39, 66, 68, 70
biological clock (*see* clock)
biological level (*see* structural level)
biological order (*see* order)
biological organization (*see* organization)
biological structure (*see* structure)
biological system (*see* system)
biological (psychological) time (*see* time)
biological unit, 47
biology (*see* theory)
bioluminescence, 103
biopolymer, 36-37, 46, 67
biosphere, 17, 73
blood pCO$_2$, 60
body, 4, 27
bond: 70
  covalent, 18, 37, 43, 66
  interatomic, 75
  non-covalent, 37

boundary: 21
  inner boundary, 47
    outer boundary, 47, 67
boundary conditions, 46
brain, 73-74, 76-77, 79, 88, 93-94, 111
brainstem centers (*see* memory centers)
brainstem reticular space, 31-32, 76, 79, 82
business organization, 61
butterfly, 102

calcium (*see* atom)
calcium oxide, 14
calendar, 25
carbohydrate, 17-18, 38-39, 70
carbon (*see* atom)
carbon compounds, 35
carbon dioxide, 13-14, 17, 70
carbon monoxide, 8
catabolism (*see* metabolism)
catabolistic system (*see* system)
catalysis, 23, 31, 43
catalyst, 23, 35, 38, 41
causal relation, 27
causality principle (*see* principle)
cell: 5, 22, 35, 37, 41, 44-47, 59-60, 63, 65,
     67, 69, 104, 108
  chloroplast, 35
  cytoplasmic gel, 21-22, 41, 45-47, 67
  cytoskeleton, 21
  division, 22, 27, 44, 48, 59, 64
  Golgi, 35
  membrane, 20, 22, 35, 42, 47-48, 65-67, 71
  mitochondrium, 35, 46, 71
  nuclear membrane, 46
  nucleus, 44, 46-47
  organelle, 35, 46
  ribosome, 41
cell surface charge, 35
central nervous system (*see* nervous system)
centrifugal force (*see* force)
change, 26, 28-29, 40, 64
character, 79
charge (*see* electrical charge)
charge of protein (*see* protein)
chemical potential, 21
chemical system (*see* system)
chemiluminescence, 103
chemosynthesis, 17

chirality, 37-39
chiral process (*see* process)
chloride ion (*see* ion)
chloroplast (*see* cell)
chromatin, 44
chromium (*see* atom)
chromophore, 100
chromosome, 44
circadian rhythm, 32, 38
circumstellar disk, 12
clay grain, 35-36
clock: 25, 30
   atomic, 30
   biological, 32
cloud:
   atmospheric, 98
   cosmic dust, 5-8, 11, 13
   hydrogen gas, 4-8
   matter, 12
   molecular, 11
   rotating, 12
   solar system, 12-13
codon, 44
cognition (*see* psychological process)
cognitive disorder, 109
cognitive level (*see* structural level)
coherent state (*see* state)
collagen fiber (*see* biopolymer)
colloid, 36
color, 78-79, 95-97, 99-102
community (*see* social group)
complexity, 8, 14, 40, 45, 60-61, 63, 71, 73, 75,
      78-79, 81-82, 89, 92, 104
concept (*see* psychological phenomenon)
cone (*see* retina)
conformation (*see* protein)
consciousness, 82, 85, 92
conservation of mass-energy, 61, 73
contemplation (*see* psychological process)
continental crust, 11-15
continents, 14
control system, 61
convection, 12
cooperative process (*see* process)
cooperativity, 18-19, 29, 39, 43, 62, 67-68, 88,
      111, 113
coordinate system, 94
coordination, 63, 67, 69, 80

core:
   earth, 11, 13-15
   galaxy, 5-8
   sun, 12
corundum, 101
cosmic dust (*see* cloud)
cosmic organization (*see* organization)
cosmology, 3, 5-6
counterions, 21
covalent bond (*see* bond)
creativity, 108-109
cultural disorder, 92, 110
cultural memory, 83
cultural order (*see* order)
culture, 83, 107-108, 111-112
crystal, 28, 47, 63, 66
cyanobacteria, 17
cycle, 62, 66-67
cyclic process (*see* process)
cyclic tendency, 62
cytoplasmic gel (*see* cell)
cytoskeleton (*see* cell)
cytosine (*see* pyrimidine)

DNA (*see* nuclei acid)
   neural, 74
dark energy, 4-5
dark matter, 4-5
Darwin adaptation, 67
day, 25, 30, 98
death, 48, 84
decay, 69
degrees of freedom, 40, 62-63
denim cloth, 100
density, 7-8, 35
depolarization, 19, 37, 40, 64, 67
development (*see* process, developmental)
diamond, 101
dielectric, 23
differentiation, 45, 60, 68, 71, 80, 108, 110
diffraction grating, 102
dipole (*see* electrical dipole)
dipole moment (*see* electrical dipole moment)
disorder, 19, 62-68, 108, 111
dissociation, 87
division (*see* cell)
dominance, 42-43, 45, 62, 64, 67
double bond, 99

dualism, 73, 82, 85-87, 91
dualistic cognitive system (*see* unveridical
    cognitive system)
dust cloud (*see* cloud)
dye, 100, 103
dynamics (*see* theory)

earth: 11, 14-15, 17-18, 22, 25, 30, 35, 39, 73,
    92-93, 98, 104, 107, 111
  axis, 25, 30, 39
  core (*see* core)
  mantle, 13
  orbit, 25
  plate, 14
  regolith, 35
  revolution, 39
  rotation, 25, 30, 39
economy, 110
electric dipole: 37, 45-46, 67, 71
  piezoelectric, 67
electric dipole moment:
  protein molecule, 22
  water molecule, 18, 22-23, 99
electrical charge, 5, 18, 21, 35-37, 46
electrical field (*see* field)
electrical polarization (*see* polarization)
electrical process (*see* process)
electrochemistry, 70
electromagnetic field (*see* field)
electromagnetic force (*see* field)
electromagnetic spectrum, 3, 79
electron (*see* particle)
electron acceptor, 18
electron donor, 18
electrostatic force (*see* force)
emerald, 101
emotion (*see* psychological phenomenon)
empirical referent, 69, 92-93, 96, 103
energy: 4-5, 7, 11, 17, 19-20, 23, 26-27, 39, 45
    69-71
  chemical, 70, 75
  electromagnetic, 75
  field, 69
  free, 37, 43, 70, 74, 83
  gravitational, 5
  heat, 13-15, 26, 43, 47, 66, 100, 103
  internal, 69
  kinetic, 19, 40, 46, 69-70, 75

  nuclear, 5, 7
  photonic, 70
  physical, 70
  potential, 69
  structural, 19, 42, 71, 99
  thermal, 7, 40, 48, 68, 75
energy conversion, 70
energy crisis, 46
energy field (*see* field)
energy level, 7, 30, 101, 103
energy organization (*see* organization)
energy-rich phosphate bond, 43
energy-transduction center, 67
entropic process (*see* process)
entropy: 26, 53-65
  negative entropy, 65
environment: 11, 17, 28-29, 35-37, 40-41, 43-45,
    47, 59-60, 63, 65, 67, 69, 73, 78-80, 88,
    91, 100
  internal, 22, 79
  socio-cultural, 83, 89, 92, 107-110, 113
enzymatic system (*see* system)
enzyme: 22, 23, 41-42, 45, 59, 65, 80
  redox, 71
epoch, 30
equilibrium state (*see* state)
ethics, 84, 113
evolution, 7, 17, 23, 30, 36, 38, 43-45, 60-61
    63, 67, 73, 75, 78, 93, 100
experience (*see* psychological phenomenon,
    concept)
explosion, 4, 103
explosion energy, 6-7

facilitation, 43, 45, 63, 65, 68, 74, 93, 108, 111
family (*see* social group)
fatty acid, 35
fear, 83, 112-113
feedback loop (*see* quantum structural
    aggregate)
feeling (*see* psychological phenomenon)
functional system (*see* system)
field:
  antigravitational, 4
  basic (universal), 3, 5, 28-29, 31, 36, 42, 60,
    69-70, 74, 108, 111
  electric, 18, 22, 37, 94, 96
  electromagnetic, 7, 36-37, 47, 94

energy, 3, 71, 104
  gravitational, 4-6, 8, 12-13, 15, 26
  helical, 39
  local ionic, 23
  local polarizing, 32, 37, 41-42, 45-46, 62, 68, 70
  magnetic, 15, 36, 94
  normalizing, 37, 43, 45, 59-60, 68, 83, 96
  resultant, 66
  unitary, 29, 31-32, 35-37, 43, 48, 64, 68, 87, 104
field asymmetry, 37, 69
field energy (see energy)
field induction, 41, 43, 45, 62, 69-71, 74, 76-77, 79, 81
field organization (see organization)
field structuring, 36
field theory (see theory)
firefly, 103
flower, 100
fluidity, 66
force: 5, 19-22, 26, 36, 40, 87
  centrifugal, 6
  electromagnetic, 74
  electrostatic, 21, 23
  intermolecular, 19, 66
  magnetic, 12, 15
  natural, 7, 30
  nuclear strong, 5-7, 74
  nuclear weak, 7, 74
  physical, 4
  repulsive, 6
  tensional, 19-21, 42-43
  van der Waals, 75
forces of nature (see force)
form: 3, 28-29, 40, 61, 63, 65, 74, 86, 93, 100, 104, 110
  symmetrical, 37-38, 40, 67
  transactional, 32
formative process (see process)
fossil, 36
fossil fuel, 60
fourth spatial dimension, 104
free energy (see energy)
free will, 87
frequency, 30, 94, 98
frequency oscillator, 30
friction, 26

fuel, 5, 7, 70-71
function, 31, 35, 37, 48, 63
fungus, 103
fusion, 7, 70

galactic time (see time)
galaxy: 3-5, 14, 29, 31, 36, 38, 104, 111
  Milky Way, 6, 11
  organic galaxy, 47
galaxy core (see core)
gas cloud (see cloud)
gas law (see law)
gas medium, 11, 13, 19-20, 98
gas pressure, 12
gel, 36, 48
gelation, 21
gelled system (see system)
gemstone, 100-101
gene, 39, 41-42, 44-45, 47, 63
gene expression, 39
genetic memory, 75, 80, 87
genetic process (see process)
genome, 44
general relativity (see theory)
geodetic surveying, 30
geometrical relation, 36
gland, 60
glass, 98
glow worm, 103
gluon (see particle)
gluthatione, 45
glycine (see amino acid)
god, 84, 92, 110-111
Golgi (see cell)
gradient (see polarizing gradient)
granite, 14
gravity (see force)
gravitational energy (see energy)
gravitational field (see field)
gravitational force (see field)
gravitational law(see law)
gravitational waves, 4
greenhouse effect, 14
Greenwich Mean Time, 30
growth, 60-61, 64, 69
ground perception, 77
guanine (see purine)

hard-problem, 79
hearing, 75, 78-79
hearing spectrum, 89
heart, 60
heat (*see* energy)
heat bath, 19
helical field(*see* field)
helium (*see* atom)
hemoglobin, 100
hereditary unit (*see* gene)
heredity, 73
heterogeneity, 62
hierarchy: 29, 39-40, 43, 45, 59, 61-63, 69,
        73, 75-76, 78, 81
    social, 83
high density water (*see* water)
histone (*see* nucleoprotein)
Homo sapiens, 60, 73, 111
homogeneity, 62
hormone, 80, 89
horizon, 98
human nature (*see* nature)
human values, 111, 113
humanity, 112-113
hummingbird, 102
hydration layer, 20-21
hydrogen (*see* atom)
hydrogen bond, 42, 44, 66, 68, 99
hydrogen gas cloud (*see* cloud)
hydrogen sulfide, 13
hydrostatic pressure (*see* pressure)
hydrothermal vents, 17
hypothalamic nuclei, 77

idea (*see* concept)
idealism, 110
imagination (*see* psychological phenomenon)
indigo, 100
individuality, 3, 40
induction (*see* field induction)
inequality of cause-effect relation, 27, 74
inflation, 4
information, 73, 82, 111
information processing, 59, 62
infrared radiation (*see* radiation)
inner boundary (*see* boundary)
inorganic process (*see* process)
insect, 100

instability, 26, 66-67
instinct, 74, 78, 83-84, 86-87, 107
intensity, 78, 103
interfacial water (*see* water)
intermolecular force (*see* force)
internal energy (*see* energy)
internal environment (*see* environment)
interpersonal relationship, 107, 111
invariant, 74
ion: 22, 23, 37, 43
    calcium, 101
    chloride, 46
    chromium, 101
    iron, 101
    magnesium, 101
    organic anion, 46
    potassium, 46
    sodium, 46
    titanium, 101
ion distribution, 22, 46
ion permeability, 46
ionic charge, 46
ionized gas, 4, 6
ionizing front, 10
iris, 98
iron (*see* atom)
irreversible process (*see* process)
isolable process (*see* process)
isolable structure (*see* structure)
isolable system (*see* system)

kinetic energy (*see* energy)
knowledge, 92, 109

l-amino acid (*see* amino acid)
language (*see* psychological process)
lava flow, 13
law:
    gas, 19, 42
    gravitational (motion), 12, 25-27, 86
    natural, 42, 61, 85, 91
    physical, 23-24, 61, 109, 110
    thermodynamical, 26, 69
learning (*see* psychological process)
length, 28
life, 7, 11-13, 15, 17-18, 20, 29-30, 35, 37, 39,
        42-43, 45, 60, 65, 67, 69-71, 84, 86-87,
        92-93, 100, 107, 111

light (*see* radiation)
light intensity, 32, 94, 96
light interaction:
   absorption, 94, 99, 102
   diffraction, 94, 101-102
   dispersion, 94-96, 98
   interference, 102
   reflection, 94-95, 98, 102
   refraction, 94-95, 98, 101-102
   scattering, 94, 97-98
light perception, 26
light spectrum, 95, 99-100, 102
light-year, 4
light wave, 94
lightening, 17
lipid, 18, 46
lipid bilayer, 35, 67
lipid membrane, 21, 46
liquid, 19, 20, 66, 98
liquid crystal, 66-68
living matter (*see* matter)
living process (*see* process)
living protein (*see* protein-ion-water system)
living structure (*see* structure, biological)
living system (*see* system, biological)
local ionic field (*see* field)
local polarizing field (*see* field)
local time (*see* time)
locomotion, 80
logic, 26
longitude, 30
low density water (*see* water)
luciferase, 103
luciferin, 103

machine, 61, 63
macromolecule, 3, 7-8, 14-15, 29, 43
magma, 13, 17
magnesium (*see* atom)
magnetic field (*see* field)
magnetic force (*see* force)
magnitude, 104
maladaptation, 92, 109
mammal, 60
manganese (*see* atom)
manipulation, 80, 108
mantle (*see* earth)
mass, 12, 28, 104

mass density, 4
mass-energy, 4-5, 12
material transport, 68
mathematics, 4, 29
matter: 4-5, 12, 26-27, 31, 73, 86, 94
   atomic, 4
   living, 20
   particle, 12
matter cloud (*see* cloud)
measurement, 61, 73, 93
mechanism, 61
medium, 26, 29
melanin, 100, 102
membrane (*see* cell)
mental memory: 31, 73-75, 78, 85, 87, 109
   central structuring centers, 75-81
   cultural, 103
   differentiation, 77, 83
   motor control centers, 75, 77, 79-81
   motor memory areas, 75-81, 88
   sensory memory areas, 75-81, 88
mental pattern (*see* pattern)
mental process (*see* psychological process)
mental structure (*see* structure)
meridian, 30
metabolism: 32, 37, 39, 70-71, 80, 100, 107
   anabolism, 29, 62, 65, 70
   catabolism, 29, 60, 62, 65, 70
metastable (*see* state)
meteorite, 13, 27
methane, 8, 13
micelle, 35
Milky Way (*see* galaxy)
mind, 31, 73, 86
minerals, 45, 100
mitochondrium (*see* cell)
molecular orbital, 99
molecular cloud (*see* cloud)
molecular vibration, 99
molecular reaction, 35
molecular structure (*see* structure)
monotheism, 84, 92
moon, 13, 15
morals, 83-84
motion, 21-23, 26, 45, 60, 69
microwave, 30
mitochondrium (*see* cell)
morphic process (*see* process)

multiplication, 60, 68
multiplication of patterns (*see* pattern)
muscular system, 80
mutation, 47

nation (*see* social group)
natural process (*see* process)
natural system (*see* system)
nature, 91, 93, 100-101, 109, 111
    human, 26, 84, 87
    physical, 3, 6-7, 26-27, 38, 63-65, 84, 87-88
negative entropy (*see* entropy)
nervous system
    autonomic-endocrine, 74, 77, 80
    central, 31, 71, 78, 92
    motor, 89
    peripheral, 29
network, 61-62
neurosis, 87
neutron (*see* particle)
Newton's laws (*see* law, gravitational)
nitrogen (*see* atom)
non-covalent bond (*see* bond)
node, 21, 61-62
normal polarization state (*see* state)
normalizing distortion, 93, 96, 109
normalizing field (*see* field)
normalizing process (*see* process)
nuclear energy (*see* energy)
nuclear fusion, 6
nuclear membrane (*see* cell)
nuclear strong force (*see* force)
nuclear weak force (*see* force)
nucleic acid, 35, 37-39, 43-45, 60
nucleoprotein, 44-45
nucleosynthesis, 7
nucleotide, 43
nucleus:
    atomic, 4, 6-7, 70
    cellular (*see* cell)
number, 60

object, 5, 38, 88, 99
ocean, 11, 13-14, 17, 35
ocean crust, 17
oligopeptide, 5
opal, 101-102
open system (*see* system)

orbit of earth (*see* earth)
order: 19, 36, 40-41, 43, 45, 47, 59, 61-68, 86,
        93, 99, 104, 110-111, 113
    biological, 44, 45, 63, 113
    distributional, 95
    orientational, 65-66, 68
    positional, 66
    social, 111-113
    spatial, 7-9
    structural, 48, 62-63, 65, 78
    temporal, 69
    unitary, 112-113
ordered region, 65-68
ordered relation, 27-28, 39-40
ordering process (*see* process)
organ system (*see* system)
organ transplantation, 60
organelle (*see* cell)
organic anion (*see* ion)
organic chemistry, 41
organic galaxy (*see* galaxy)
organism: 18, 25-27, 29, 31, 43, 45-47, 60-63,
        65, 67-70, 73, 78, 87, 91-92, 100, 104,
        108, 111
    unicellular, 17, 47, 59
    multicellular, 47, 60, 67
    thinking, 61
organism-environment system (*see* system)
organization:
    atomic, 7
    biological, 5, 28-29, 42, 45, 48, 60-62, 65,
        69, 80, 110
    cosmic, 26
    energy, 7, 59
    field, 36
    protoplasmic, 26
    self-organization, 35-36, 59
    structural, 5, 31, 36, 43, 45-46, 59, 63
    universal, 5
orientational order (*see* order)
oscillations, 18, 23, 67-68
osmosis, 20
osmotic pressure (*see* pressure)
outer boundary (*see* boundary)
outer shell, 31
outgassing, 13
ovum, 68
oxidation, 46, 70

oxidation-reduction, 45
oxidative phosphorylation, 70
oxygen (*see* atom)
ozone, 15, 17

particle: 4-5, 28-29, 40, 46, 69, 94, 100
  boson, 82
  electron, 4-5, 7, 17-18, 30-31, 63, 70-71
  gluon, 4-5, 30
  photon, 4, 15, 30, 94, 96-97, 99, 101-103
  proton, 4-5, 7, 15, 63
  quark, 4-5, 30
  neutron, 7
  primary, 31, 38, 74
particle physics, 4
parts to whole relation, 39-41, 62, 67-68
past experience (*see* concept)
pattern:
  atomic, 40, 68
  dominant, 41
  mental, 31, 74-75
  social, 107
  spatial, 74
  stable, 41-42, 65-66
  structural, 29, 31-32, 39-42, 45, 61, 63, 65,
    69, 88-89, 91, 93-94, 96, 109-110
  unstable, 41
pattern configuration, 45
pattern extension, 41
pattern multiplication, 41-42, 65
peacock, 102
perception (*see* psychological process)
perceptual level (*see* structural level)
permanence, 26, 87
personality, 83, 92, 107-112
personality trait (*see* psychological
    phenomenon)
pH, 22
phase transition, 66, 68-69
philosophical systems, 83-86, 111-112
phosphate, 35, 43
phosphorus (*see* atom)
photon (*see* particle)
photoreceptor, 32
photosynthesis, 17, 70, 100
physical force (*see* force)
physical nature (*see* nature)
physical network, 61

physical universe (*see* universe)
physics:
  one-way (*see* theory, unitary physics)
  reversible (*see* theory, classical physics)
physiological level (*see* structural level)
piezoelectric dipole (*see* electric dipole)
pigment, 100-101
Planck time, 23, 31
plane, 37-38
planet, 4, 7, 25-26, 45, 98
planet orbit, 26
planetary motion, 32
planetesimal, 13
plant, 17, 39, 100
plasma, 5-6
polar asymmetry (*see* structural asymmetry)
polar molecule, 23
polarized system (*see* system)
polarizability, 19, 36
polarization, 18, 36-37, 39, 42, 45-47, 59,
    66-67, 70
polarization pulses, 48
polarized system (*see* system)
polarized structure (*see* structure)
polarizing gradient, 37, 46-47
polymerization, 19
polysaccharide (*see* biopolymer)
posture, 80
potassium ion (*see* ion)
potential difference (*see* polarizing gradient)
potential energy (*see* energy)
pressure: 7, 13, 19-22
  hydrostatic, 18
  osmotic, 20
pressure pixel, 19, 42
primary particle (*see* particle)
principle:
  causality, 27
  symmetry, 27
  unitary, 6, 27-28, 31, 41-42, 62, 79
prism, 95, 102
probability, 64
process: 3-8, 12, 14, 17, 28-29, 35, 39-42, 60, 62,
    64-65, 67-68, 71, 73, 111
  adaptive, 36, 60, 65, 73, 80, 113
  autocatalytic, 41, 74
  biochemical, 75
  chiral, 39, 64

cooperative, 18, 68-69
cyclic, 45, 60, 65-66, 70, 74
developmental, 29-36, 39, 41-43, 45, 59, 64-
    65, 68-69, 71, 74-75, 79, 87, 91, 104, 111
electrical, 26
entropic, 64-65
formative (structural), 39, 41-43, 60-61, 71,
    75-77, 82-83, 85, 90-91, 98, 105-110
genetic,73
inorganic, 63
irreversible (one-way), 26-28, 74, 87
isolable, 27-28, 65, 69
mechanical, 112
living, 43, 65, 111
mental (*see* psychological process)
morphic, 65
natural, 25, 30
normalizing, 29, 32, 37, 41-43, 45-47, 62-63,
    67-69, 76, 80-81, 93, 108, 110-112
ordering, 28, 32, 87
radiative, 26
reversible, 26
self-limiting, 40
social, 112
unitary, 28-31, 41, 44-45, 59, 64, 69, 73, 76,
    78, 91, 104, 111
vectorial, 62, 64, 67
propaganda, 110
protein: 15, 18, 20, 22, 36-38, 41-44, 59, 63, 66-
    67, 69
    charge, 22, 44
    cluster, 20
    conformation, 23, 38, 45
    domain, 20-21, 37, 42
    dynamics, 37
    filament, 21
    folding, 22
    hydrophilic region, 21
    hydrophobic region, 21
    pattern, 59-60
    surface, 43
    structure, 36, 38, 45
protein-gel phase transition, 22
protein-ion-water system (*see* system)
proteoglycan (*see* biopolymer)
proton (*see* particle)
proton gradient, 71
protoplanetarium disk, 12

protoplasmic organization (*see* organization)
protosun, 12
pulse of depolarization-repolarization, 65-66,
    68, 71
psychological phenomenon: 75, 76, 78
    abstract concept, 80-81, 84, 88-89, 93, 109
    concept, 61, 63, 65, 76, 80, 86-88, 93, 111,
        113
    behavior, 76-78, 81, 83-84, 92, 107, 109-110,
        113
    emotion, 76, 79, 83, 89, 107
    feeling, 79, 89
    percept, 31, 80
    personality trait, 81
    thought, 29, 61, 76, 78, 81, 83, 84, 87, 93,
        111
psychological process: 83, 86-87, 91-92, 108, 113
    affection, 79
    anticipation, 31
    attention, 81, 83
    conception (cognition), 79, 83, 91-92, 109
    contemplation, 81-82
    imagination, 73, 81, 92
    judgment, 89, 110, 113
    language, 75, 79-80, 88-89, 107, 108-109
    learning, 65, 77, 80, 84, 110, 112
    meaning, 81, 89
    perception, 32, 76, 78-80, 86, 88, 91-92, 109
    reasoning, 81, 92
    thinking, 80-81, 83, 87-88, 92, 109, 113
psychological whole, 76
pulsation, 37
pupil, 95, 103
purine: 35, 44
    adenine, 43
    guanine, 44
pyrimidine: 33, 44
    cytosine, 44
    thymine, 44

quality, 61, 110-111
quantitative method, 110-111
quantitative relation, 61
quantity, 28, 61, 86, 92-93, 110-113
quantum, 42
quantum field (*see* theory)
quantum field particle, 31-32, 46, 63
quantum mechanics (*see* theory)

quantum overtone transitions, 99
quantum structural aggregate: 31-32, 76, 78, 81-82, 103
  central process, 81
  configuration, 79-80
  feedback loop, 77-79
  sequence, 32, 80, 81, 92
  serial coupling, 76, 79
quark (*see* particle)
quasar, 4, 6

RNA (*see* nucleic acid)
rain, 13, 15, 108
rainbow, 95, 98, 102
radiation: 8, 12, 15, 17
  infrared, 14
  light, 5, 26, 79, 93-94, 100, 102-103
  solar, 12, 17, 93, 100
  ultraviolet, 11, 15, 17, 100
  radioactivity, 13, 26
radiative process (*see* process)
radical, 95
radio astronomy, 8
reality, 25, 28, 36, 40, 84, 91, 93-95, 104, 109
realm:
  inorganic, 6, 68
  organic, 15, 31, 59, 65, 68, 71
receptor: 75, 96
  external, 75
  internal, 75, 79
  proprioceptor, 75, 77, 79
  smooth muscle, 75
  skin, 75
receptor cell, 14
record, 41, 45, 74, 76
redox enzyme (*see* enzyme)
red giant, 8
regolith (*see* earth)
regulation, 61
reification, 109
refied concept, 110
reindeer, 103
relativity (*see* theory)
relaxation (*see* depolarization)
religious systems, 82-84, 111-112
repolarization, 37
reproduction, 46, 87
repulsive force (*see* force)

repulsive gravitational field (*see* field)
residual asymmetry (*see* structural asymmetry)
resonance, 7
resonance frequency, 30
respiratory depth, 60
respiratory system (*see* system)
resultant field (*see* field)
retina: 95, 96
  cone, 95, 96
  rhodopsin, 95
  rod, 57, 96
reticular space, 31-32, 76, 79, 82
reversibility, 27
reversible process (*see* process)
reversible system (*see* system)
revolution (*see* earth)
rhodopsin (*see* retina)
rhythmicity, 32
ribosome (*see* cell)
rod (see retina)
rose, 100
rotation, 12, 36, 38
rotation of earth (*see* earth)
rotational time (*see* time)
rotating cloud (*see* cloud)
ruby, 101

sapphire, 101
science, 111-112
scientific method, 61, 93 , 109-110
sea, 17, 20, 60
secretion, 68
self-facilitation, 108
self-limiting process (*see* process)
self-organization (*see* organization)
self-regulation, 68
semipermeable membrane, 20
sensory stimulus, 75
sensory system, 85-91
signaling pathway, 67
silica, 14, 102
silicate, 8, 13-14
silicon (*see* atom)
social groups: 107, 109-110, 113
  community (society), 83, 107, 111-113
  family, 107
  nation, 107, 113
social disorder, 112-113

social environment (*see* environment)
social order (*see* order)
social structural aggregates, 107-108
sodium (*see* atom)
sodium ion (*see* ion)
sodium-potassium pump, 46
solar mass, 4
solar radiation (*see* radiation)
solar spectrum (*see* light spectrum)
solar system (*see* system)
solar system cloud (*see* cloud)
solar wind, 15
solid, 19, 42, 68
solubility (*see* water)
solute, 18, 20-21
solute cluster, 21, 42
solution, 20
solvent, 21, 42
soul, 83-84, 92
sound, 78, 79, 89, 103
sound reception, 103
sky, 98
skylight, 99
sleep, 39
smell, 75
surface charge (*see* electrical charge)
space:
   interstitial, 22
   intracellular, 22
   intravascular, 22
   physical, 4-8, 13, 15, 18, 25, 27, 29
      35-37, 39, 94, 104
spacetime, 25
spatial order (*see* order)
spatial pattern (*see* pattern)
spatial symmetry, 27
special relativity (*see* theory)
spectroscopy, 8
spin, 12, 15, 39, 104
spin axis, 104
spin state (*see* state)
spirit, 73, 84
spring, 25, 37
stability, 7, 37-38, 40-42, 59, 62-63, 65-68, 70
stable system (*see* systems)
standard time (*see* time)
star, 3-8, 11-12, 18, 29, 36, 69, 91-92, 98, 104
state: 27, 28, 41
   asymmetrical, 70
   coherent, 32
   cyclic, 69
   equilibrium, 37, 62, 64-65, 67, 69
   excited, 99
   ground, 70
   metastable, 38
   normal polarization (norm), 68
   spin, 70
   stable (symmetrical), 41, 64, 69-70
static concept, 84
statistical mechanics (*see* theory)
statistical quantity, 64
stimulus, 37, 65-66, 74-57, 77-78, 93
stratosphere, 17
strong force (*see* force)
structural asymmetry: 27-29, 31, 37-38, 40-44, 45-
      46, 62-65, 67-71, 74, 76, 78, 80, 94, 96
   axial, 36
   polar, 36-37, 46, 68
   residual, 41
structural energy (*see* energy)
structural organization (*see* organization)
structural (organizational) level:
   biochemical, 59, 62
   biological, 60
   cognitive, 60, 62, 75, 77, 79, 81
   perceptual, 62, 75, 78
   physiological (tropistic), 62, 75, 78-79
   social, 108
structural order (*see* order)
structural organization (*see* organization)
structural pattern (*see* pattern)
structural quanta, 42
structural transformation, 29-30, 40, 42-43,
      45, 62, 64, 66, 68-70, 76, 82, 89, 92
structure: 3, 7, 18, 19, 26-28, 35-37, 40-41, 43,
      61-62, 69
   atomic, 7, 31, 74, 94
   biological (organic), 14-15, 23, 28, 30-31, 35-
      36, 40, 42, 45, 61-66, 68, 70, 87-88, 93-
      94, 96, 104
   cosmic, 64
   cyclic, 69
   helical, 39
   isolable, 40
   mental, 28, 93-94
   molecular, 19, 31, 74

quantum field, 45, 74-75
  physical, 14-15, 30, 40, 63
  polarized, 29, 39, 41, 46, 68
  stable, 47, 70
  wave, 18
substrate, 60
succession relation, 25, 27, 94
sulfides, 17
sulfur (*see* atom)
sun, 4, 6, 11-12, 15, 45, 93, 97
sun core (*see* core)
sunlight, 12-15, 17, 46, 70, 93, 97, 99
sun orbit, 12
sunrays, 17, 94, 97, 100
sunrise, 98
sunset, 98
superposition, 32
superstition, 83, 111
supper massive density, 4-5
symbol, 88, 108
symbolic concepts, 80, 109, 111
symmetrical form (*see* form)
symmetrical relations, 27
symmetry: 27-29, 31, 62-63, 69-70
  bilateral, 38
  point-centered, 70
  structural, 32, 39-43, 45, 64, 67, 68
  rotational, 28
  translational, 28, 41, 70
  uniaxial, 66
symmetry principle (*see* principle)
synthesis, 7, 35, 41, 68, 70
  basic, 39, 59
  organic, 41-42, 45, 59, 70, 76
system: 12, 26-27, 29, 36-37, 40-41, 64, 69, 74
  biological, 3, 11, 22, 28, 30-31, 35-37, 39-40,
    46, 61-62, 64-65, 67-70, 75-76, 92
  catabolistic, 45
  complex, 26, 40, 62-63
  chemical, 3
  enzymatic, 69
  functional, 71
  gelled, 45
  irreversible, 26
  isolable, 29, 31, 39, 41-42, 60, 69
  mental (*see* psychological system)
  natural, 25

open, 36
organ, 60
organism-in-environment, 47, 60, 64, 111
  polarized, 46
  physical, 3, 30-31
  protein-ion-water, 37, 67-68
  respiratory, 71
  reversible, 12
  self-constructed, 61
  solar, 4, 11-13, 15, 25-26, 97
  stable, 65
system behavior, 39-40, 75

tact, 89
taste, 75
technology, 110, 112-113
telescope, 3, 104
temperature, 5-8, 11, 13, 17, 22, 23, 66
temporal order (*see* order)
temporal relation (*see* asymmetrical relation)
tendency (*see* unitary tendency)
tension (*see* force)
tetrahedron, 18, 38
thermal energy (*see* energy)
thermodynamics (*see* law)
thermonuclear fusion, 5
thermoregulation, 100
theory:
  atomic, 28
  biology, 19, 26, 40, 61-63, 66, 70, 73, 87
  classical dynamics, 26, 39-40, 63
  classical physics, 25-27, 36, 42, 61, 65, 87,
    94, 108
  electrical, 46
  exact science, 28, 67
  fluctuon model, 87
  information, 75
  process, 28, 40, 42, 73
  quantum field, 4, 20, 32, 47, 75, 87
  quantum mechanics, 20, 73, 87, 94, 99
  general relativity, 5, 6, 25, 87
  special relativity, 25
  statistical mechanics, 19
  statistical thermodynamics, 64
  string, 25
  thermodynamics, 27, 37, 63-64
  unitary, 28-29, 32, 39-43, 46, 62-64, 67-71,
    73-74, 78, 88, 91, 94, 108

thermal energy (*see* energy)
thermodynamics (*see* laws)
thermonuclear bomb, 7
thermonuclear fusion, 5
thermoregulation, 59
thermostat, 22
thinking organism (*see* organism)
thought (*see* psychological phenomenon)
thought development, 83
threshold, 6, 65, 68, 81
thymine (see pyrimidine)
time: 4-7, 12, 19, 23, 25, 27-28, 30-31, 35, 39,
    45, 59, 73
  astronomical, 30
  atomic, 30
  biological (psychological), 27, 30-32
  galactic, 5-6
  irreversible, 25-26
  local, 30
  physical, 31, 103
  mean solar, 30
  relativistic, 26
  reversible, 25-26
  rotational, 30
  standard, 30
time arrow, 25-26, 31
time fundamental unit, 30
time paradox, 26
time perception (*see* psychological
    phenomenon)
time SI unit, 30
time zone, 30
tissue: 31
  conductive, 65
  contractile, 65
touch, 79
tradition, 107, 109
transactional form (*see* form)
tropistic level (*see* structural level)
two-valued orientation, 110

ultraviolet radiation (*see* radiation)
unconscious, 81
uniaxial symmetry (*see* symmetry)
unicellular organism (*see* organism)
uniform region, 65
uniqueness, 3
unitary field (*see* field)

unitary principle (*see* principle)
unitary order (*see* order)
unitary process (*see* process)
unitary tendency, 28-29, 31, 41, 43, 62, 65-66,
    68, 76, 82, 88, 94, 104, 107-108
universal field (*see* field)
universal organization (*see* organization)
universe, 3-7, 11, 25-27, 29-30, 36, 91, 104,
    107, 111
unstable pattern (*see* pattern)
unveridical cognitive system, 108-112
unveridical concept, 92, 108

valence electron, 99, 103
valine (*see* amino acid)
vascular system, 60
vectorial process (*see* process)
velocity, 12, 18, 23, 31, 62
veridical cognitive system, 113
veridical concept, 92, 111
vesicle, 35
vibration, 19, 67
vicinal water (*see* water, interfacial)
visual system, 75, 78-79, 89, 91, 95-96, 104
vitalism, 87
volcano, 13
volume, 19-20

water: 7-8, 11, 13-15, 17-20, 23, 35, 37, 59, 68,
    71
  bulk, 21-22
  cluster, 19-21, 42-43, 66, 68, 71
  density, 21-22
  deuterized, 99
  droplet, 97
  high density water, 43
  low density water, 43
  molecule, 19-21
  network, 19, 21-22, 31
  interfacial, 21-22
  particle, 19
  phase transitions, 22
  photolysis, 70
  quantum, 19
  solubility, 22-23
  structure, 18, 20-23, 31, 43, 71
  vapor, 13, 99
  wave structure, 18, 20-21

wave, 46, 94
wavelength, 20, 94-96, 100, 102
wave structure (*see* water)
weak nuclear force (*see* force)
whole, 59, 62, 64-65, 67, 69-70
word, 78-79, 83, 88-89
work, 70-71, 88
world: 7
  verbal, 88, 109
  inorganic, 101
  macroscopical, 63
  mesoscopical, 21
  organic, 101-102, 104
  physical, 7, 23, 29-31, 35, 91-92, 94, 104,
      109
  social, 108